SEALS and SEALERS

A Pictorial History of the Newfoundland Seal Fishery

Based on the Cater Andrews Collection

SHANNON RYAN
Assisted by Martha Drake

BREAKWATER

Canadian Cataloguing in Publication Data

Ryan, Shannon, 1941-
Seals and sealers

Includes index.
ISBN 0-920911-19-6

1. Sealing – Newfoundland – History – Pictorial works. 2. Sealers (Persons) – Newfoundland – Pictorial works. 3. Andrews, Cater – Photograph collections. I. Drake, Martha. II. Andrews, Cater. III. Title

SH362.R95 1987 639'.29'09718 C87-093044-3

Breakwater gratefully acknowledges the support of The Canada Council.

The Publisher acknowledges the financial contribution of the Cultural Affairs Division of the Department of Culture, Recreation and Youth, Government of Newfoundland and Labrador which has helped make this publication possible.

Foreword

Let us take the road from Newtown through Pound Cove and Templeman, Wesleyville and Brookfield, Badger's Quay and Valleyfield, Pool's Island and on to Greenspond. As we follow the sinuosities of the shoreline, the sea appears now on the right, now on the left, while our road crosses tickles and inlets, skirts the coves, and links the coastal fringe of low-lying islands with the rocky main. Nowhere else in Newfoundland are the twin elements of rock and sea more intricately compounded to make a landscape.

On a bright summer day, the blue sea, the pink granite, and the tiny patches of incredibly green garden and lawn occupying the hollows between rocky outcrops make a crazy quilt of startling beauty. In winter, when Labrador current and northeast wind press northern ice upon the shore, only the line of ballicater marks the place where land ends and the world of Arctic floes commences.

Here, in the heart of Bonavista North, Cater Andrews was born and raised. Bred in his very bone and marrow was a love of this land and a pride in the strength and courage of the people who had built upon the rock and who had learned to draw their livelihood from the resources that the sea and the running ice brought to them.

In such an environment no one could be untouched by the living epic of the great annual seal fishery. The common stock of song and story told of heroic deeds, of acts of derring-do, of high adventure, of inspired leadership, of courage in the face of terrible dangers, of battles for survival against fearful odds; and, on the darker side, of callous exploitation, of elemental forces of destruction, of tragedy, of death. Nor were the characters in those tales remote story-book figures. The great sealing skippers, the Keans, the Barbours, the Winsors, the Hills, and others whose names were familiar household words throughout Newfoundland, were, in Wesleyville, living, breathing, everyday members of the community. And so too were their crews: the men who sailed the ships and braved the sea and the ice, the men who were the stuff of which legends were made. Who could resist the fascination of this world in which strong men and splendid ships pitted their strength and their courage against the forces of sea and ice? What youth would not aspire to have his manhood proved in such an arena?

Even when Cater Andrews had become a quiet and scholarly academic, presiding over the biology department of Memorial University of Newfoundland, he retained his fascination for this world in which he had been nurtured. Throughout his life he maintained his contacts with his home and his people. And he kept a dream: that one day he would see published a definitive work that would tell the world the real story of the seals and the sealers.

To this end he laboured incessantly, taking every spare minute from his official duties that his health and strength would allow to collect and compile the voluminous materials upon which his books would be based. His conception was broad, his collections all encompassing. No aspect of the subject would be untouched, no facet unexplored. From the natural history of the seals and the physical and ecological environments in which they lived, through the history of their exploitation and its impact on the social and economic development of Newfoundland, to the lore of ships and men that formed the ethos of a community, he would preserve once and for all the record of this remarkable Newfoundland phenomenon.

Regrettably, his health failed and he died before his great task could be completed. His surviving spouse, Martha Iona Hill Andrews, herself the daughter of one of the great sealing captains, recognizing the worth of the remarkable archive that her husband had assembled, and desiring to see his work continue, donated the collection to the University and to the care of the literary executors of the estate. The executors, for their part, undertook to have the collection properly housed, protected, indexed and catalogued, and to make it accessible to worthy scholars committed to the realization of the vision that had brought it into being.

As the work of cataloguing has progressed it has become apparent that within the collection there is a remarkable aggregation of photographs depicting virtually every phase and aspect of the seal fishery. Inasmuch as these represent a fascinating pictorial record of great historical and social significance, the executors have determined that a selection of them should be published at the earliest possible time.

This volume is the result, and we commend it to all those desirous of closer acquaintance with one of the dominant themes of Newfoundland's traditional culture as portrayed through some remarkable examples of the photographer's art.

Leslie Harris
for The Literary Executors
The Cater Andrews Collection

January 1987

Preface

This study of Newfoundland's spring seal fishery owes a tremendous debt of gratitude to the late Dr. Cater Andrews who collected material on this subject throughout his adult life. Following his death in 1978 his wife, Mrs. Martha Andrews, very generously presented his collection to Memorial University of Newfoundland where it is presently housed under the supervision of three literary executors: Dr. Frederick Aldrich, Dr. Leslie Harris, and Dr. George Story.

I would like to thank Mrs. Andrews and her daughter Mrs. Judy Mallam; the literary executors; and Mr. Richard Ellis, University Librarian, for making it possible for me to have access to the Cater Andrews Collection.

In addition, I would like to thank Dr. Harris and the University Administration for making financial support available to identify, organize and catalogue the many photographs in this Collection. It was this project, with additional support from the University, which led to the production of this photographic history.

In preparing the book every effort has been made to identify, obtain all relevant information, and give appropriate credit for each photograph. (Notice of any errors or omissions in this regard will be gratefully received and the appropriate corrections made in any subsequent editions.)

Special thanks are due to the following people who helped identify photographs, provided technical advice, and aided with the general research:

Captain Carl Barbour, St. John's, Newfoundland; Mr. Fred Bragg, retired sealer, St. John's; Mrs. Gwen Brokenshire, St. John's; Mr. Manny Buchheit, photographer, St. John's; Mr. Sean Finlay, Editor, *The Evening Telegram*, St. John's; Ms. Judy Foote and staff, University Relations, Memorial University of Newfoundland; Captain James Gillett, retired, St. John's; Ms. Nancy Grenville, Archivist , Centre for Newfoundland Studies (CNS), Memorial University; Ms. Anne Hart, Librarian, CNS, Memorial University; Mr. Philip Hiscock, Archivist, Memorial University Folklore and Language Archives; Mr. Tony Murphy, Archivist, Provincial Archives of Newfoundland and Labrador; Professor Gerry Panting, Chairman, Maritime Studies Research Unit, Memorial University; Mr. William Parsons, Curator, Conception Bay Museum, Harbour Grace, Newfoundland; Miss Evelyn Power, Photograph Archivist, *The Evening Telegram*, St. John's; Dr. Bobbie Robertson, Office Secretary, Newfoundland Historical Society, St. John's; Mr. Andrew Short, retired sealer, Riverhead, Harbour Grace; Mr. Gordon Thorne, retired sealer, St. John's; Ms. Heather Wareham, Archivist, Maritime History Archives, Memorial University.

As well, I would like to thank Memorial University of Newfoundland for their permission to publish photographs from the Cater Andrews Collection. Only pictures from this Collection have been included in this book. (There were several poor quality photographs in the Collection which had been reproduced from clippings from *The Evening Telegram*. In these few cases, *The Evening Telegram* very generously allowed me to refer to their files and to use their original photographs and negatives.)

I would like to say a special thank you to those who assisted in the final physical preparation of this book: Mr. Jack Martin, Photography, Memorial University, who prepared most of the photographs for publication; Ms. Fran Warren, who typed the manuscript; Mrs. Debbie Foley and Ms. Cathy Horan, who proofread it; Mr. Paul Kenney, who prepared the index; and the management and staff of Breakwater Books who with their assistance and patience have made the publication of this book possible.

Finally, I would like to thank Mrs. Martha Drake who assisted me with every phase of this book. Mrs. Drake was introduced to the Cater Andrews Collection in September 1986 when she was hired to identify, organize and catalogue the photographs in the Collection. She quickly became expert in the field of seal fishery photography and adept at retrieving information from the Collection.

Mrs. Drake also did most of the travelling associated with the identification of the photographs and the production of the manuscript, and dealt directly with many of the people whose assistance is acknowledged above. In this respect, Mrs. Drake would like to add her thanks to mine, and she would like to thank especially her young son, Andrew, for his understanding and support.

It has been a rare privilege to work on the Cater Andrews Collection and it is an honour and a pleasure to dedicate this book to the memory of the man who created it, Dr. Cater W. Andrews of Wesleyville.

S.R.

Introduction

Newfoundland's origin as a colony and its later economic, cultural and political development rested upon and was shaped by two principal fishing industries: the cod fishery which drew European fishermen to Newfoundland and the seal fishery which made it possible for them to stay.

Beginning about 1500 European fishermen came to Newfoundland annually to catch codfish for consumption back home. Most of these fishermen fished on the Grand Banks of Newfoundland and filled their ships with salted cod. The English preferred to live on the island and dry the salted cod in the sun before bringing it back to Europe where it was sold as *saltfish*. This inclination to operate from shore stations was one of the factors that encouraged the development of permanent fishing premises on Newfoundland's northeast coast. Although the establishments were permanent, few of the people who lived in them during the winter months could be classified as permanent inhabitants. English fishermen interested in emigrating to the New World usually moved to New England; most of the others commuted back and forth across the Atlantic between the fishing grounds and their homes in England. Only a few wintered on the island on a regular basis.

By 1715, after two hundred years of constant European contact, Newfoundland had about three thousand permanent English residents and these people had become an integral part of the large annual migratory fishery with its headquarters in the southwest counties of England, especially Devon and Dorset. These inhabitants produced saltfish during the summer months, sold their product to the visiting fleet in exchange for the basic necessities and worked during the autumn and winter months building fishing boats and maintaining the large premises of the fishing fleets. They could grow very little food except cabbage and turnips and, after the 1750s, potatoes. Consequently the number of inhabitants remained low.

During the eighteenth century the number of English inhabitants in Newfoundland increased. Wars encouraged fishermen to remain on the island because they feared capture at sea by enemy ships and impressment at home by their own navy. The introduction of the potato in 1750 aided this development by providing an important source of food and vitamin C. Furthermore, the British conquest of New France and the American Revolution both encouraged Newfoundland shipping and population growth. By 1750 the island had about six thousand permanent residents, and by 1800, twenty thousand.

The French Revolutionary and Napoleonic Wars brought major changes to Newfoundland. The demand for saltfish in Europe increased and this encouraged more fishermen to come to the island. Meanwhile, the magnitude of the war in the north Atlantic and in northern Europe induced them to stay. The market for saltfish was practically unlimited and the traditional fishing harbours — such as Harbour Grace, Carbonear and St. John's — became overcrowded for the size of their fishing grounds and drying facilities. Consequently the local ship owners began sending fishing crews to northern Newfoundland harbours and to the southern Labrador coast for the summer cod fishery. This was the beginning of the annual migration of many of the island's fishermen to the Labrador coast for the fishing season. The development of this cod fishery on the northern coast and later on the Labrador, which was prosecuted by small ships from the more heavily populated harbours to the south, particularly in Conception Bay, coincided with the rapid growth of another industry — the seal fishery.

Harp seals had always been plentiful in the waters around Newfoundland but only on a seasonal basis and never in the summertime. Consequently the visiting English fishermen were never in a position to exploit the vast herds of this sea mammal. By 1700, however, the inhabitants had begun to net a few seals near headlands and to shoot them in open water. Occasionally, when the ice-floes carrying seals would drift close to shore everybody in the vicinity would club or shoot as many as possible. The seal meat was consumed locally while the pelt, with its thick layer of fat, was used to produce oil and leather. The oil was used for lighting and for various local purposes, with the surplus product being sold to the British ships. Similarly, some leather was used to meet local needs and the rest was exported. It was found that the farther north one went, the better one's chances were to harvest seals; therefore, the gradual spread of settlement into the northern harbours in the eighteenth century was connected with the growing awareness of seal resources. Statistics can be found for ten years between 1723 and 1735 and from them it can be seen that the total value of seal oil exported from Newfoundland averaged £3111 (sterling) annually. During the ten-year period ending in 1795 the annual average was valued at £7345 (sterling). Almost all of this production came from the northern bays.

By the beginning of the 1800s the fishermen from Conception Bay and St. John's had begun to go out in small ships to look for seals in the spring of the year. Gradually these expeditions discovered that the harp seals bore their young on the ice-floes around the last of February. This discovery led to the development of Newfoundland's second-most important fishery — the spring seal fishery.

Fishermen discovered that if they could reach the harp seal herds or *patches* during early March, when the females had their young, they could kill a large number of these *whitecoats*. It also became evident that most of the same men and ships employed by Conception Bay and St. John's firms in the northern Labrador cod fishery from June to September could be used in the seal fishery from February to May with little increase in investment. Therefore, the Labrador cod fishery and the spring seal fishery complemented one another from the beginning.

The seal fishery expanded rapidly as the fishermen became acquainted with the habits of this sea mammal and as they acquired the knowledge needed to judge correctly the location of the *main patch* and the expertise necessary to take their little fishing ships through the unpredictable ice-floes . During the Napoleonic War period about 100,000 seals were killed annually; in 1818 over 165,000 were killed, and in 1822 over 386,000; total annual production rose to over 600,000 seals in 1831 and to about 700,000 during peak years in the 1840s.

The seal fishery was stimulated by the demand for oil in Great Britain, and seal oil exports began to rival saltfish exports during the early nineteenth century. In 1815 about 1400 tuns of seal oil were exported (one tun equals about 1125 litres). Exports rose as the seal harvest increased and during the early 1830s over 7500 tuns of seal oil were exported annually. Throughout most of the 1830s and '40s about one-third of the total value of Newfoundland's exports was derived from seal oil. (The less important seal skins were also exported for use in Great Britain's leather industries.) Similarly, on average, about one-third of the fishermen's income was derived directly from the spring seal fishery. From February to May men were employed preparing ships and supplies, then actually killing and pelting and bringing the catch successfully to port, and finally processing the pelts. The latter activity involved separating the fat from the skins for the production of oil and semi-finished leather. This operation required premises with *seal-skinning* facilities and large vats. Local cooperages provided the many casks and drums needed for the shipment of the oil and skins to market. By June most of the work involved was over, and then all attention was directed to the prosecution of the cod fisheries.

This revolutionary development in the colony's economy during the 1820s, '30s and '40s resulted in a substantial population increase and other demographic changes. The centres which had initiated the seal fishery were there to obtain the early benefits: St. John's and the Conception Bay ports of Harbour Grace, Carbonear and Brigus prospered and expanded. The old cod fishing establishments were redesigned and expanded to accommodate the new industry. In 1827 there were 290 ships and 5418 men engaged in the seal fishery (excluding those involved in processing). Investment increased until 1857 when 370 substantially larger ships and 13,600 men participated. The population of the island rose from 40,000 in 1815 to 124,000 in 1857, a growth unmatched by any other similar period in Newfoundland's history. The population of St. John's increased from 12,000 in 1815 to 30,000 in 1857, while that of Conception Bay increased from 12,000 to 33,000 during the same period. Meanwhile, during this period of growth it was discovered that bigger ships were more productive. This led to a gradual increase in the average tonnage of the sealing fleet and consequently a consolidation of the industry in the hands of the larger firms, one which could afford this increasingly greater investment. However, despite the larger investment in men and ships, the peak harvests of the 1840s could not be repeated in the 1850s owing to overexploitation of the seal herds.

The seal fishery, like the Arctic whale fishery, had always suffered from the inadequacies of sail power. Sailing ships could move through ice-fields only when the wind was favourable and even then they lacked the heavy power to force a track and had to rely totally on open leads of water. On the open sea sailing ships were very efficient freight carriers as the clippers and the windjammers were to demonstrate for decades to come, but in a field of ice they were powerless. Consequently, the early adoption of the steam engine and the screw propeller by the British Arctic whalers in the 1850s caught the immediate attention of the larger Newfoundland sealing/fishing companies and they were quick to follow suit.

Two St. John's firms — Walter Grieve & Co. and Baine Johnston & Co. Ltd. — were the first in Newfoundland to acquire steamers for the seal fishery. In 1863 the former company sent the SS **Wolf** to the ice while the latter sent the SS **Bloodhound I**. With nominal horse power (nhp) of thirty and forty, respectively, these *wooden-wall* steamers were low-powered and bore little resemblance to the twentieth century steel steamer. The first wooden walls carried full sails and their steam-driven engines were used more as a source of auxiliary power. However, the maneuverability of the steamers was their most attractive and useful feature. They could steam directly to the *Front* off Newfoundland's northeast coast unhindered by the prevailing westerly and northwesterly winds. Once in the ice-field, the steamers could steam in any direction through the loose ice, although they were very easily jammed when the ice tightened. Walter Grieve & Co. and Baine Johnston & Co. Ltd. proved the value of the early steamers and the combined total of their small steam fleets had increased to four ships by 1865. In 1866 Bowring Brothers in St. John's acquired its first steamer, the SS **Hawk**, followed by Job Brothers & Co., also of St. John's, with the SS **Nimrod** in 1867. In 1866 Ridley & Co. of Harbour Grace acquired the SS **Retriever** and in 1867 added the SS **Mastiff**, while Munn & Co., also from Harbour Grace, acquired the SS **Commodore** in 1871. No other outport took advantage of the new steam technology and within a few years only Harbour Grace and St. John's were active centres of the spring seal fishery.

As steamers were introduced into the seal fishery the use of sailing ships quickly declined. In 1867 Harbour Grace sent fifty sailing ships with 2504 men to the ice, while in 1886 this port sent four steamers and 898 men and no sailing ships. In 1866 St. John's sent 177 sailing vessels and five steamers; in 1880 its fleet consisted of twenty-four steamers carrying 5089 men. A few small sailing crafts from Bonavista and Notre Dame Bays continued to engage in the seal fishery but they were a component of the annual *landsmen* seal fishery in this area and therefore totally dependent on the seals that could be found close to shore. The sailing fleet which had always gone in search of the *whelping ice* and the *main patch* of whitecoats was replaced by a smaller fleet of steamers within a few years.

This change from sail to steam coincided with the decline in the value of the seal fishery to the Newfoundland economy. The herds had been overexploited and by the end of the nineteenth century average annual production had declined to about 250,000 seal pelts. At the same time, the value of seal oil decreased as petroleum products took away much of the market for seal oil and other traditional oils. Seal oil prices declined from about $200 per tun in the early 1860s to an average of $133.10 annually during the 1870s, $104.30 during the 1880s, and $77.30 during the 1890s. In the 1830s and '40s, as has been shown, the seal fishery made up about one-third the total annual value of Newfoundland's exports. In the 1850s the proportion had dropped to about twenty-four percent. During the 1890s, while Newfoundland's total economy was depressed, the seal product exports were worth, on average, only nine percent of the total value of the colony's exports. There is no doubt that by 1900 the seal fishery had lost its commanding role in the Newfoundland economy.

The changing nature of the seal fishery in the latter nineteenth century resulted in several developments. The outports lost their sealing industries when they did not, and probably could not, replace their sailing fleets with steamers. While the inhabitants from these outports could travel to Harbour Grace or St. John's for *berths* to the ice, the smaller outports lost many of the auxiliary industries and the jobs they had created (i.e. ship building, seal skinning and processing, cooperages). Then, in 1894 the firm of Munn & Co., Harbour Grace, went bankrupt and its sealing fleet was taken over by Baine Johnston & Co., St. John's. Although Munn's premises continued to be used for some years afterwards, the entire sealing fleet was now centred in St John's. The decline in the seal fishery and its transfer to St. John's resulted in a decline in the Labrador cod fishery. On its own, this industry could not support the ships and ports which had grown dependent on seal pelts and Labrador codfish.

Thus the population of the old sealing centres also declined. Harbour Grace's population dropped from 14,727 people in 1884 to 12,671 in 1901, while Carbonear's declined from 6,206 to 5,024 and Port de Grave's from 8,698 to 7,445 during the same period.

Employment on the docks and fishing ships of Boston replaced the seal fishery in the lives of most of the people of Conception Bay. Besides the large numbers who decided to remain in Boston permanently, many more migrated annually between their homes in Newfoundland and their jobs in the United States. Nevertheless, in its reduced form the seal fishery from St. John's continued as an important industry and at times during the early 1900s it gave every indication that, in the long term, its role in Newfoundland's economy, although diminished, would remain intact.

Meanwhile, steamship technology was improving all the time. The use of sail declined as engines became more powerful and some of the older surviving wooden steamers underwent refit as masts were shortened, yard-arms removed, and engines improved. In 1906 Harvey & Co. sent the first iron-clad steamer, the **Adventure** to the ice-fields and this experiment was so successful that the company invested in two others, the **Bellaventure** and the **Bonaventure**. Other companies were quick to follow suit. Soon the iron-clad steamers **Beothic**, **Nascopie**, **Fogota**, **Sagona**, **Seal**, **Sable Island**, **Florizel** and **Stephano** were added to the St. John's fleet and, as Chafe described it, "From 1906 to 1914 Newfoundland had the finest fleet of Sealers and Ice-Breakers in the world."[1]

Chafe thought that the iron-clad steamers set the pace for the wooden walls and, as a result, the captains of the latter pushed their ships beyond capacity. Perhaps because of this a considerable number of wooden-wall steamers were lost while prosecuting the seal fishery during this period: the **Greenland** and the **Leopard** in 1907; the **Panther**, the **Walrus**, and the **Grand Lake** in 1908; the **Vanguard** and the **Virginia Lake** in 1909; the **Iceland** in 1910; the **Labrador** in 1913; and the **Kite** and the **Southern Cross** in 1914.

Because it was becoming increasingly difficult to operate the expensive iron-clad steamers profitably in a seasonal fishery that lasted only one or two months, most were sold abroad during World War I; the SS **Stephano**, the largest at 2143 net tons, was torpedoed and sunk in 1916, and the second largest, the SS **Florizel** ran aground and was likewise a total loss. At the end of the war the fleet consisted of old wooden walls and several iron-clad steamers with some of the latter being chartered from the passenger and mail services just for the spring sealing seasons. During the 1920s and '30s the annual sealing fleet continued to dwindle in numbers despite the addition of the SS **Beothic** (1926-40), the SS **Ungava** (1928-40), the famous SS **Imogene** (1929-40), and the return of the SS **Nascopie** (1927-1930).

Bowring Brothers was the principal firm still prosecuting the seal fishery during this period and their acquiring the SS **Imogene** was the final major investment in a genuine sealing steamer. The **Imogene**, 715 net tons, 379 nhp, was not the largest sealer but she was specially built and powered to cope with Arctic ice. She was a narrow ship and therefore not of much use as an ice-breaker because her track, or wake, was not wide enough for most big ships. However, the **Imogene** could get to the seals and acquire a load without difficulty. After her loss in a 1940s incident unrelated to sealing she was never replaced.

Newfoundland's sealing steamers, both wooden-wall and iron-clad, brought problems of profitable utilization to their owners. The seal fishery provided employment for only one or two months and it was often a major effort to find other ways to keep each ship in use. In the early days of sailing vessels there was never any problem because the vessels went to the Labrador cod fishery during the summer and some of the larger ones transported saltfish to markets overseas. The steamers were not suited to the cod fishery on the Labrador coast, although one

[1]Levi Chafe, *History of the Newfoundland Seal Fishery* (St. John's: 1923), p. 26.

or two were usually used to transport men and their families and supplies to that coast. Others were used in the colony's mail and passenger service along the coast. However, the wooden wall steamers were most in demand for summer work in Arctic (and on a few occasions Antarctic) waters. They were hired by the Hudson's Bay Co. to bring supplies to the Company's northern posts; they were chartered for exploration work by Shackleton, Peary and Scott, among others; and they were used by the Canadian government for survey work and government business. (Sealing captains like Bob Bartlett often accompanied the ships.) However, in the twentieth century Arctic work declined in volume and in any case the old wooden walls were too inefficient to be used. They were generally left tied up in St. John's harbour while attempts, not always successful, were made to find remunerative employment for the iron-clad steamers. Meanwhile the changing fortunes of Newfoundland during World War II encouraged many people to accept full-time employment on the new military bases in the colony. In addition, there was a general change in Newfoundland's economic orientation, and consequently, as steam sealing ships left the industry they were not replaced.

The end of the steam sealing fleet came in the 1940s. In 1941 only four old wooden walls — the SS **Neptune**, **Terra Nova**, **Eagle** and **Ranger** — went to the ice. In 1950 the **Eagle**, the last of the fleet, was intentionally sunk by her owners.

Besides the changes in the sealing ships throughout the nineteenth and twentieth centuries, the place of the spring seal fishery in the lives of the fishermen changed as well. In the beginning nearly all fishermen on the northeast coast went to the ice. They were considered co-investors and everybody paid a sum towards fitting out and supplying the ship for the seal fishery. Later this investment evolved into a set fee or charge which the ship owner required from each fisherman/sealer. Then, as the sealers began to view themselves as employees of the ship owners, they agitated for the elimination of this fee. Strikes in 1842, 1860, and 1902 eventually eliminated this *berth money*, but after 1902 sealers were allowed to purchase from the ship owner a $9 *crop* on credit. The owners deducted $12 from the share of each sealer who took advantage of this credit. However, if the share amounted to less than $9 the owners bore the loss. Moreover, sealers on the sailing ships in the nineteenth century shared one-half of the proceeds of the catch while the other half went to the ship owner. When steamers entered the industry only one-third of the catch was shared out among the men. Men accepted this primarily because the steamers were much more productive than the sailing ships and consequently each man's share on a steamer was generally larger than it had been on the sailing ships.

Meanwhile, the seal fishery continued to decline. By the 1920s the annual average catch amounted to 145,000 pelts and the number of men employed had declined to less than 2,000 from a high of between 12,000 and 13,000 in the 1850s. At the same time the focus of manpower and expertise shifted from Conception Bay and St. John's to Bonavista North. Captains like Azariah Munden, Henry Thomey and Richard Pike from Conception Bay were replaced by the Barbours, Keans and Winsors of Bonavista North. The vacuum created by the decline of Conception Bay's Labrador cod fishery was filled by the captains and men of Bonavista North's more stable Labrador cod fishery.

By the 1920s and '30s the seal fishery provided the opportunity for only a few men to earn a little money. However, demand always exceeded supply and it was difficult to obtain berths. Local shopkeepers usually had a few berths for favoured customers or for customers who owed them money; clergymen, likewise, often had a few to distribute; and the captain and officers and owners disposed of the rest.

Life as a sealer on the eve of World War II was not much different than it had been at the beginning of the steamer seal fishery in the 1860s. Men were divided into teams or *watches*, usually four per ship and these were placed under the control of a master watch who was responsible to the second hand, as the first mate was called, although the captain often dealt directly with the master watches. The master watches led their men on the ice and assigned and supervised their other duties, particularly gathering ice, shifting coal and stowing pelts.

The men's clothing and footwear had not changed much during the decades and their workday was as long as ever. They rose about 4:00 a.m., dressed in their sealing clothes which usually consisted of high boots with taps and heels studded with sharp *sparbles* and *chisels*, heavy trousers and shirts and sweaters over heavy underwear. They also had oil clothes which they wore or carried.

They went on the ice about 5:00 a.m., killed seals with their iron-tipped gaffs and pelted their take. Then they either towed the pelts to designated pans of ice where they were picked up by the steamer or, if the steamer were jammed in the ice, towed them all the way — sometimes for miles — back to the ship. They went back on board the steamer at 7:00 or 8:00 p.m. Then there was the *watch on* or going *on watch* for half the men from 8:00 to 11:00 p.m. or 8:00 p.m. to midnight. These would come off *watch* at 11:00 or 12:00, eat and go to their bunks while the other half came *on watch* from 11:00 p.m. to 3:00 a.m. or midnight to 4:00 a.m., depending on the custom aboard the steamer. While their *watch was on* men gathered ice from nearby pinnacles for drinking water and any nearby ice for covering the pelts; they shifted coal from the pounds in the hold up to the bunkers near the engine room, and they stowed the cooled pelts into the pounds covering each layer with ice. By 4:00 a.m. it was time for breakfast and another day on the ice.

Men were spared the task of handling canvas sails as their ancestors had done on board the nineteenth century sailing ships. However, from the men's point of view the constant work involved in shifting hundreds of tons of coal in baskets by hand was not a great improvement. Needless to say, the sealers put in an exhausting day and were too tired to worry about the state of their sleeping quarters or the quality of their food.

Their bunks were temporary wooden affairs built between decks and usually made passable with the addition of the sealers' own homemade mattresses which were large brin bags filled with wood shavings or hay. Coal dust, fat and blood permeated their living quarters, bunks and clothing as they always had. At night those sleeping were usually exposed to the open sky and the elements as the hatches were taken off and coal was raised up from the lower hold, past the men's bunks, to the deck. Then the seal pelts were dropped down into the hold. It was not unusual for lumps of coal and greasy, bloody seal pelts to fall on those asleep in the bunks.

The food, and the attention paid to food, changed a little during the decades. In the days of the small sailing ships men often lived on hardbread and raw salt pork and drank the semi-saltfree water which accumulated in hollows on the ice. When steamers came on the scene it was easier for men to cook some food because of the availability of coal. Men usually cooked their own food — boiled or fried salt pork, fried seal hearts and livers, hardbread and tea — in their 'spare time'. In the twentieth century the owners provided cooks to prepare the men's food, although some sealers continued to fry meals of seal hearts and liver and onions in their 'spare time' late at night. The regular food left much to be desired. Breakfast was usually *lop scouse* which consisted of a watery stew-like mixture. The sealers collected this in their tin pans and went back to their sleeping quarters where they ate/drank it with hardbread and tea. On the ice the men ate a lunch of hardbread and anything else they could afford to bring with them from home — usually a mixture of rolled oats and raisins. They drank a mixture of molasses and ice-water and sometimes added a little patent medicine to it. The men often scooped out a little hole in the snow, put in rolled oats and added a little molasses and/or patent medicine, stirred the mixture which was further diluted and thinned with the melting snow, and then ate it. The main meal back on board late in the evening was usually a piece of salt pork, a piece of boiled duff (a mixture of flour and water cooked in a bag to a hard, tough consistency) with molasses poured over the lot. One out of every three or four men usually collected the food from the galley in his boat's kettle and brought it back to the bunk area and shared it with his *mess*. By the 1930s soft bread and butter were distributed, in small quantities, fairly regularly.

The men supplied their own boat's kettle, tin pans, mugs and cutlery. They also were required to bring their own knives and steels (for sharpening the knives). Gaff handles and iron gaff points were provided by the ship owners, but the men put them together. The hauling ropes were also provided by the owners and the men spliced the ends according to the common practice.

By and large, by 1940 men were still living and working in nineteenth century conditions on ships which were no longer useful nor acceptable for any other enterprise.

During the 1940s a number of diesel-powered ships with cleaner conditions (no more coal dust!) began to operate, some from Halifax. Bowring Brothers withdrew from the industry when its steamers ceased operating and smaller operations replaced it. Landsmen and motor vessels continued to take seals, but in fewer numbers than ever. By the 1970s quotas of 120,000 to 150,000 seals were commonly set by government. Then, in the early 1980s, markets for whitecoat pelts could no longer be found, motor vessels stopped participating, and sealing reverted to what it had been in the eighteenth century — a small, widely distributed, landsmen operation.

The annual Newfoundland spring seal fishery was an enterprise of epic proportions. Ships were no match for the ice-floes if the ice tightened and some ships were crushed in this way. More commonly rudders and propellers were damaged and ships had to be abandoned. Others, like the SS **Southern Cross**, sank while en route to or from the ice-fields. The SS **Viking** accidentally blew up. Similarly, men were often caught in snowstorms and froze or drowned helplessly. The crew of the SS **Greenland**, for example, found themselves cut off from their ship (itself jammed in ice) by the sudden appearance of a large *lake*. Forty-eight men died. The crew of the SS **Newfoundland** spent about fifty-three consecutive hours in a snowstorm, unable to find their ship, and seventy-seven men were lost. Ships were lost and men died but while the industry was profitable it flourished. Finally the lack of profitability lead to its demise.

The Newfoundland seal fishery was unique. Nowhere else had a sizeable community of people become so dependent on this particular sea mammal. The seal fishery served Newfoundland well and its effects on our folklore and culture can still be seen.

1

A patch of harp seals (*Phoca groenlandica*) on whelping ice. The sealing industry depended upon *patches* of seals like this. When the seal fishery was at its peak of production during the 1840s, between 400,000 and 700,000 seals were killed annually by Newfoundland fishermen. The pelts were processed in the colony and the fat and skins were exported to the oil and leather markets of Great Britain.

2

The *whitecoats*, or young harp pups, provided the Newfoundland fishermen with their chief supply of seal pelts.

3

The adult harp seal was also slaughtered for its pelt, but in fewer numbers than the young harps.

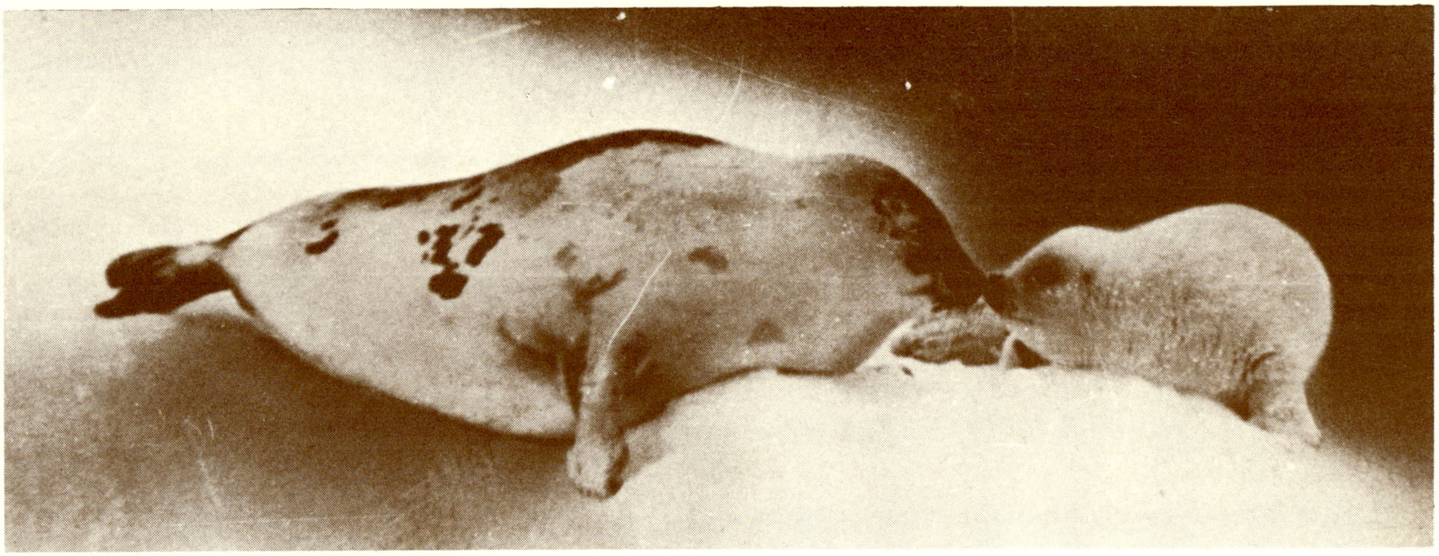

4

A female harp and her pup.

5

A mother seal watches its pup on the edge of the whelping ice.

6

A whitecoat nursing.

7

Groups of adult harp seals were often found in small lakes of water among the ice-floes. Shooting seals in these *swatches* of water was an important component of the seal fishery.

8

Hood seals (*Cystophora cristata*) were found in family groups consisting of the young *bluebacks*, the old *bitch* and *dog*. Harvesting hoods was a smaller component of the Newfoundland seal fishery.

9

From these photographs one can appreciate the size and physical characteristics of the whitecoat. The man on the left was probably Mr. J.C. Grouchy, a medical officer on the SS **Eagle** in 1934. He is wearing jodhpurs and an aviator's cap, an expensive style of attire at the time. The unidentified sealer on the right is wearing the more typical sealers' garb.

10

During the nineteenth century Harbour Grace was the most important outport sealing centre; its economy was so dependent on the Labrador cod fishery and the spring seal fishery that the town declined as these industries declined. However, in the 1850s the outport had reached its peak and this photograph, one of the oldest sealing photographs, shows the Harbour Grace sailing fleet leaving for the ice sometime during that decade.

11

St. John's, the colony's capital, became an early and prominent participant in the seal fishery. This sketch illustrates the departure of the port's sailing fleet for the ice-fields circa 1860. It was not unusual for harbours on Newfoundland's north and east coasts to freeze over in the winter, and often the departure of the sealing fleet marked the beginning of each port's shipping season.

12

The SS **Retriever**, a fully-rigged, low-powered steamer of 237.95 net tons, 50 nominal horse power (nhp), was the first sealing steamer to be employed by a Harbour Grace firm. In 1866 the **Retriever**, under the command of Captain James Murphy, was sent by Ridley & Co. to the ice-fields where she produced a record catch. With only 105 men, the **Retriever** brought in 17,450 seal pelts on her first trip and 5,950 on her second; her crew made $303 each. She was sent to the seal fishery by Ridley & Co. up to 1872 when she was lost while sealing off Cape St. Lewis, Labrador. Her captain, Richard Pike, and the crew were rescued by the SS **Mastiff**.

In this painting men are depicted killing, pelting, and then towing the pelts to the ship. In the days of the sailing ships it was the practice to tow the pelts to the ships as shown; as steamers became more powerful, however, the seal pelts were usually placed in piles and the ship steamed through the ice-floes hoisting them aboard.

13

The SS **Lion**, 292.42 net tons, 75 nhp, was one of the first wooden-wall steamers to be used in the Newfoundland seal fishery. (The first steamers, the **Bloodhound I** and the **Wolf**, went in 1863.) As this painting illustrates, these early wooden-wall steamers often carried full sails. Supplied by Walter Grieve & Co., she made her first voyage to the ice-fields in 1867. With such masters as Francis Ash and Alexander Graham, the **Lion** went to the annual seal fishery for fifteen years and brought in a total of 170,125 seal pelts.

While returning from the seal fishery in 1870 under Captain Graham, the **Lion** flew a blue ensign with large white letters that read "No Confederation." In 1871 the **Lion** brought home the crew of the **Wolf** after that steamer had been cut in two by an iceberg in Green Bay. The **Lion** left St. John's on 6 January 1882, a bright and calm moonlit night, bound for Trinity under Captain Patrick Fowlow. She was never seen again, although a small amount of debris was found near Baccalieu Island. The prevailing explanation was that her boilers had not contained enough water and, as a result, the ship was blown to pieces.

Grandfather, age 19, made 3 trips on Walrus in 1889; earned £28.

14

The SS **Walrus**, 183.53 net tons, 60 nhp, was brought from Scotland in 1870 by J. & W. Stewart for the Newfoundland seal fishery. She was 145 feet long, 25 feet 8 inches wide, and 12 feet 4 inches deep, and carried a large spread of canvas. Between 1870 and 1908 the **Walrus** missed only one spring at the ice — that of 1892 — and brought in 291,259 seal pelts in all. In 1880 she set a record for the shortest successful voyage to the ice, leaving Greenspond under the command of Captain Joseph Barbour on March 10 and arriving back fully loaded on March 21. In 1908 the **Walrus** was severely damaged butting through the ice and when the ice slackened on April 3, she filled and sank; Captain Jacob Winsor and the crew were picked up by the other ships in the area.

In this photograph, taken by Ern Maunder, the **Walrus** is landing seal pelts on the southside of St. John's.

15

The SS **Hector**, 290.64 net tons, 70 nhp, was built in 1870 as a whaler/sealer by Alexander Stephen & Sons, Dundee, for Job Brothers, Liverpool. She was 151.1 feet long, 24.1 feet wide and 16.6 feet deep. Job Brothers & Co. Ltd., St. John's, sent her to the seal fishery for the first time in 1871. That year, under the command of Edward White Sr., she made three trips to the seal fishery and brought in 27,096 seal pelts, the largest number ever brought in on one vessel up to that time. The crew received $109.50 each. In 1891 she was sent to Scotland, rebuilt, rechristened, and returned to Newfoundland as the SS **Diana**. She continued to prosecute the seal fishery for Job Brothers and, after 1915, for James Baird Ltd. As the **Hector**, she brought in 200,040 seal pelts, and as the **Diana**, an additional 348,929, for a total of 548,969 pelts, making her the fifth-most productive sealer in Newfoundland's history. As the **Hector** she was commanded by Edward White Sr., Edward White Jr., William Knight, William Barbour (who continued to command her when she became the **Diana**), James Barbour and, for an eleven-year period beginning in 1898, by Alph Barbour. A number of different masters then took her to the ice until her final voyage in 1922. Besides her career as a sealer, the **Diana** operated as a whaler, on Arctic expeditions and on the Newfoundland-Labrador mail service. In 1922, during her fifty-second consecutive spring, the **Diana** became jammed in the ice at the Front and lost her tail shaft. She was abandoned on March 27 and sank while burning.

In this photograph the **Diana**, under the command of Captain William Barbour, arrives in St. John's from the seal fishery, 25 April 1892.

16

The SS **Iceland**, 287.31 net tons, 70 nhp, was a sealer/whaler built by Alexander Stephen & Sons, Dundee. She was 150.5 feet long, 27.3 feet wide, and 16.4 feet deep. She became one of Nicholas Stabb & Sons' sealers and went to the ice every spring from 1872 to 1880. The **Iceland** then was purchased by John Munn & Co., Harbour Grace, and for the next fifteen years was one of four steamers going regularly from that port (the others being the SS **Greenland**, the SS **Mastiff** and the SS **Vanguard**). The Harbour Grace fleet ceased to operate after 1895 and these steamers became a part of Baine Johnston & Co. Ltd.'s fleet. Among the captains of the **Iceland** were Samuel Blandford, Charles Dawe, Darius Blandford and William Winsor Sr. Her total catch, brought into Harbour Grace and St. John's, amounted to 420,121 seal pelts. On 30 March 1910 the **Iceland** was crushed by ice at the seal fishery off the Funk Islands and her crew was rescued by the SS **Eagle**.

This photograph, taken in Harbour Grace by James Vey, shows Harbour Grace South in the background.

17

The SS **Greenland**, 259.11 net tons, 75 nhp, pictured here in the foreground, was brought from Aberdeen in 1872 by Nicholas Stabb & Sons, St. John's. She was sent to the ice annually by Stabb until 1881 when she became a part of the small Harbour Grace fleet supplied by John Munn & Co. After Munn's bankruptcy, ownership passed to Baine Johnston & Co. Ltd. During her career, under such captains as George Barbour, Isaac Barbour and Charles Dawe, she brought in 395,668 seal pelts. The **Greenland** continued to engage in the seal fishery until 1907 when, after breaking her shaft while sealing off the island's east coast, she was crushed by ice in a gale. This wooden steamer, however, is best known for the tragic loss of forty-eight of her men on 21-22 March 1898 during what became known as the **Greenland** disaster.

This 1888 photograph, taken at Harbour Grace by Reuben T. Parsons, shows the SS **Vanguard** in the background.

18

The SS **Vanguard**, 322.72 net tons, 85 nhp, was built in Scotland in 1872 for John Munn & Co., Harbour Grace. She was 166 feet long, 28 feet wide, and 18 feet deep. Under the command of Captain Azariah Munden, the **Vanguard** went to the ice for the first time in 1873, and then went on to become Munn's most important sealer. When ownership passed to Baine Johnston & Co. Ltd. in 1895 she continued sealing under such captains as Henry Dawe, Charles Dawe, George Barbour, Darius Blandford and William Winsor Sr. During her career in Newfoundland she brought in a total of 394,864 seal pelts.

Having rescued the crew of the SS **Resolute**, lost at the seal fishery in 1886, and the crew of the SS **Grand Lake**, which sank at the ice in 1908, on 11 April 1909 the **Vanguard** broke her main shaft and sank while at the seal fishery. Her men were rescued by the SS **Algerine**.

This photograph by James Vey shows the **Vanguard** in solid ice in St. John's harbour, circa 1900.

19

The Harbour Grace waterfront showing the premises of R.D. McRae & Sons (formerly Ridley & Co.) on the left. The sealing steamers **Vanguard**, **Iceland**, and **Greenland** were formerly owned by John Munn & Co., Harbour Grace, but with that company's bankruptcy in 1894 these sealers became part of Baine Johnston & Co. Ltd.'s sealing fleet. Thus, after one hundred years as a major sealing centre, Harbour Grace lost its fleet to St. John's, marking the end of direct outport participation in the steamer seal fishery.

Photograph by S.H. Parsons, circa 1895.

20

The SS **Neptune**, 465.19 net tons, 110 nhp, a wooden whaler/sealer, was built in 1872-73 by Alexander Stephen & Sons, Dundee. She was 190.5 feet long, 28.9 feet wide and 15 feet deep. Her construction was overseen by Captain Edward White Sr., her master for the following ten years. Brought to Newfoundland for Job Brothers & Co. Ltd., St. John's, the **Neptune** prosecuted the seal fishery for sixty-seven springs, was commanded by twelve different captains, and brought in a total of 1,223,691 seal pelts. In 1873 she was the largest of the seventeen ships at the seal fishery, and in 1888 she set a record for seal pelts brought in by a wooden steamer — 42,242. Ownership of the **Neptune** passed to the Ungava Steamship Co. in 1934 and back to Job Brothers in 1943. That same year she sank near St. John's with a load of coal.

This photograph shows the **Neptune** as she was after her refit, circa 1900, when her main mast and her yard-arms were removed and the two other masts shortened.

21

The SS **Proteus**, 467.36 net tons, 110 nhp, was a wooden whaler/sealer built in 1873 by Alexander Stephen & Sons, Dundee, for J. & W. Stewart, St. John's. She was 190.4 feet long, 29.9 feet wide, and 18.6 feet deep. A sister ship to the SS **Bear** (see caption 23), the **Proteus** was built for work in the ice: she had an iron prow, extra sheathing on the sides, and a spare rudder and propeller. In 1874 she was sent to the seal fishery under the command of Captain Pierre Mullowney, and from 1875 to 1883 she was commanded by Captain Richard Pike of Harbour Grace. In ten years the **Proteus** brought in 139,146 seal pelts, setting a record in 1875 by bringing in 44,337 in two trips to the ice.

In 1881 the **Proteus**, under the command of Captain Pike, was chartered by the United States Congress to transport the Greely expedition and enough provisions for two years to Lady Franklin Bay in the Arctic, with the agreement that she would return for them two years later. On her return in 1883, again under the command of Captain Pike, the **Proteus** was crushed in the ice at Smith Sound on July 23. Captain Pike and his men walked six hundred miles over the ice to safety.

In this S.H. Parsons' photograph, the **Proteus** is stuck in the ice in Harbour Grace.

22

The SS **Leopard**, 217.28 net tons, 60 nhp, was built in 1873 at Ayre, Scotland, for Walter Grieve & Co. She was engaged in the Newfoundland seal fishery for thirty-three consecutive years and brought in a total of 248,816 pelts. Her captains included Levi Diamond, George Hiscock, Henry Dawe, George Hann, Job Kean, Peter Blackwood, and finally, Bob Bartlett. In 1907, while en route to the seal fishery in the Gulf under the command of Captain Bartlett, the **Leopard** ran ashore near Ferryland. The crew ripped up the deck sheathing to make a road over the ice to the shore a short distance away, enabling all the men to get ashore with their clothes before the **Leopard** sank.

Photograph by S.H. Parsons, 1901.

23

The SS **Bear**, 468.27 net tons, 110 nhp, a steam sealer/whaler, was built in 1874 by Alexander Stephen & Sons, Dundee, for Walter Grieve & Co. She was 190.4 feet long, 29.9 feet wide, and 18.6 feet deep. She was used in the Newfoundland seal fishery for ten consecutive years, bringing in a total of 125,916 seal pelts under such captains as Alexander Graham, Levi Diamond and Francis Ash.

In 1884 the **Bear** was sold to the United States government and, along with the SS **Thetis**, was sent to the Arctic to rescue the American explorer, Greely. She was then transferred to the U.S. Coast Guard Service and served on the Alaska patrol until 1929 when she was purchased by the city of Oakland, California and converted into a maritime museum. It was at this time that she was used in the filming of Jack London's novel, *The Sea Wolf*. She was used by Admiral Richard E. Byrd on his second Antarctic expedition between 1933 and 1935, and again on his 1939-41 expedition. During the Second World War the **Bear** returned to coast guard duty, and in 1947 was purchased by a Canadian company, which apparently intended to reconvert her to a sealer. If this had been the actual plan, it was abandoned, and she was laid up in Nova Scotia until she was purchased by Philadelphia interests with plans to convert her into a museum and restaurant near that city. While being towed there in 1963, she sank in a storm.

24

The SS **Thetis**, 491.67 net tons (reclassified to 396.19 tons in 1916 and 279.01 in 1924), 98 nhp, was built in 1881 by Alexander Stephen & Sons, Dundee. She was 181 feet long, 30.5 feet wide, and 19 feet deep. One of the most powerful of the early steamers, from 1881 to 1884 the **Thetis** was used in the Newfoundland seal fishery by William Stephen & Co. In 1884 she was bought by American interests and, with the SS **Bear**, was sent to the Arctic to bring back the Greely expedition. Brought back to Newfoundland by Job Brothers & Co. Ltd. in time for the 1917 seal fishery, she was used regularly in the seal fishery by Job Brothers and later by the Ungava Steamship Co. She was employed as well on the Newfoundland-Labrador mail service.

During her career, the **Thetis** brought in 293,861 seal pelts under Captains Charles Dawe and Alexander Fairweather in the 1880s and later under W.C. Winsor, Peter Carter, William Bartlett Sr., Samuel Wilcox, John Dominey, Jacob Kean and Abram Kean. After being damaged in the ice during the spring of 1936, the **Thetis** returned to St. John's, discharged her cargo, and continued to leak so badly she was towed outside the harbour and beached.

In this photograph, the **Thetis** is leaving for the seal fishery on 10 March 1920.

25

The SS **Ranger**, 353.62 net tons, 70 nhp, was built in 1871 by Alexander Stephen & Sons, Dundee, for J. & W. Stewart. She was 161 feet long, 28.7 feet wide, and 18 feet deep. J. & W. Stewart employed her in the seal fishery annually until 1894 when her ownership was transferred to A.S. Rendell & Co; C.F. Bennett & Co. sent her to the seal fishery in 1902; and in 1903 Bowring Brothers became her owners. She remained with this firm for the rest of her career. The **Ranger** also served on the Newfoundland-Labrador coastal service.

Beginning in 1872 (with the exception of 1915), and commanded by seventeen different captains, the **Ranger** prosecuted the seal fishery annually up to and including 1941, spending a record number of sixty-eight springs at the seal fishery, and bringing in a total of 934,007 seal pelts — the second largest number brought in by any sealer. During a nine-year period from 1882 to 1890, under the command of Captain Joseph Barbour, the **Ranger** brought in an impressive 219,776 seal pelts.

The **Ranger** was damaged at the ice-fields on a number of occasions and was practically rebuilt in 1917. In 1942, during her sixty-ninth spring at the ice, she was damaged in a storm and sank while her crew made it to safety.

In this photograph by J.C. Grouchy the **Ranger** is preparing to leave Bowring Brothers' wharf for the 1935 seal fishery.

26

The SS **Bloodhound II**, 376.26 net tons (reclassified later to 313.93 net tons), 96 nhp, was built in 1872 by Alexander Stephen and Sons, Dundee. She was 160 feet long, 32.2 feet wide and 18.3 feet deep. She was sent to the seal fishery by Baine Johnston & Co. Ltd. for two springs, 1873 and 1874, and brought in a total of 33,038 seal pelts.

The **Bloodhound** was then purchased by the British government, renamed the **Discovery**, and sent on an Arctic relief expedition. Upon her return to England she was used as a navy store ship and finally put up for sale.

Baine Johnston & Co. Ltd. purchased her in 1903 and returned her to the Newfoundland seal fishery in 1904, again with the name **Bloodhound**. In the meantime, her net tonnage had been altered to 313.93, and she was reportedly in excellent condition and well suited for the seal fishery. In 1904 one writer described her as "the ship of the fleet" for strength and durability. Between 1904 and 1917 the **Bloodhound** was sent to the seal fishery every spring and during this fourteen-year period brought in a total of 147,764 seal pelts. Her commanders included Alfred Smith, George Barbour, Darius Blandford, William Winsor, Jacob Winsor, George Clarke, Alph Barbour and George Whiteley. On 15 July 1917, while en route to St. John's, the **Bloodhound** went ashore and was lost at Point La Haye, St. Mary's Bay.

27

The SS **Aurora**, 367 net tons, 98 nhp, was built in 1876 as a whaler/sealer by Alexander Stephen and Sons, Dundee, for William Stephen & Co. She was 165.2 feet long, 30.6 feet wide and 18.9 feet deep. She was one of the first Dundee ships to prosecute the Newfoundland seal fishery on a regular basis, beginning in 1877. Ownership was transferred to Bowring Brothers in 1894 and she continued to prosecute the seal fishery until she was sold abroad in 1911 and went on to be used in Antarctic explorations by Sir Ernest Shackleton, among others.

During her thirty-five years at the Newfoundland seal fishery the **Aurora** brought in a total of 587,374 seal pelts. Prominent among her masters were James Fairweather, Arthur Jackman, Abram Kean and Daniel Green. In June 1917 the **Aurora** left Wellington, New Zealand en route to England and disappeared with all hands.

This photograph, taken by James Vey circa 1900, shows the **Aurora** in St. John's harbour.

28

The SS **Viking**, 275.73 net tons, 86 nhp, was a wooden whaler/sealer built in Norway in 1881. During her Norwegian career she was best known as the ship on which the famous Norwegian explorer Fridtjof Nansen made his first voyage north in 1882. The **Viking** was purchased by Bowring Brothers in 1904, and made at least twenty trips to the Gulf seal fishery under the command of Captain William Bartlett Sr., but spent her final voyages under several sealing captains at the Front. She brought in a total of 270,341 seal pelts under Captains William Bartlett Sr., William Bartlett Jr., George Whiteley, Isaac Barbour and Richard Badcock.

In 1931, while under the command of Captain Abram Kean Jr., the **Viking** left for the Front carrying, besides her complement of sealers and crew, Varick Frissell and his movie production team, who were putting the final touches to their film, *The Viking*. On March 15, while near the Horse Islands, the **Viking** was torn apart by an explosion and sank with the loss of twenty-four men, including Frissell.

29

The SS **Terra Nova**, 450.28 net tons, 120 nhp, was a wooden whaler/sealer built in 1884 by Alexander Stephen and Sons, Dundee. She was 187 feet long, 31 feet wide, and 19 feet deep. Beginning in 1885, the **Terra Nova** prosecuted the seal fishery for William Stephen & Co., St. John's; in 1894 ownership passed to the Dundee Seal & Whale Fishing Co.; and in 1898 she was purchased by Bowring Brothers. The **Terra Nova** continued to go to the seal fishery every year until 1903, when she was bought by the British Admiralty and sent to relieve Captain Robert Scott's SS **Discovery** in the Antarctic. She was repurchased by Bowring Brothers in 1906 and sold again in 1910, when she returned to the Antarctic with Scott's second expedition. On her return to Newfoundland in 1913 she was once again purchased by Bowring Brothers and, from 1914 to 1942, prosecuted the seal fishery every spring.

The **Terra Nova**, the biggest whaling ship of her time, was one of the longest-serving and most productive wooden walls in the Newfoundland seal fishery: during fifty-one springs at the Newfoundland seal fishery she brought in over 850,000 seal pelts, the third-largest number of any sealer. Among her captains were Arthur Fairweather, Charles Dawe, Arthur Jackman, Abram Kean, John Parsons and Stanley Barbour. In September 1943 the **Terra Nova** sprang a leak off southwest Greenland and sank without loss of life.

In this photograph, listed as "courtesy of H.G. Posting," the **Terra Nova** is using steam and sail power in pack ice, probably in the Antarctic. Even by the beginning of this century the wooden-wall steamers often carried a full complement of sails.

30

The SS **Erik**, 412 net tons (reclassified in 1908 to 461.62 net tons), 80 nhp, was a whaler/sealer built in 1865 by Alexander Stephen and Sons, Dundee, for G. Gibbs & Sons. Constructed of wood with iron plating, she measured 157.8 feet long, 29.5 feet wide, and 19.5 feet deep. After prosecuting the seal and whale fisheries from Scotland, in 1882 the **Erik** was purchased by the Hudson's Bay Co. and served in the Arctic for twenty years. The **Erik** then was brought to Newfoundland and sent to the seal fishery by Job Brothers & Co. Ltd. from 1902 to 1914 when ownership was transferred to James Baird Ltd. She continued as a sealer until the end of the 1918 season. Her captains included Job Kean (for twelve years), Nathan Kean and John Parsons, and during her seventeen-year career she brought in a total of 212,406 seal pelts.

On 24 August 1918 the **Erik** came under attack by a German submarine off St. Pierre. Before sending her to the bottom, the U-boat transferred the crew to a nearby fishing schooner.

This photograph shows the **Erik** on dry dock in 1908, probably as a result of her collision with the SS **Aurora** on March 16 of that year.

31

The SS **Newfoundland**, at 567.83 net tons, 130 nhp was the largest wooden wall. Built around 1872, she was purchased by Captain J. Farquhar and his associates in 1893 and outfitted for the seal fishery, which she prosecuted under Captain Farquhar, out of Nova Scotia in 1893 and 1894. In fact, the St. John's *Evening Telegram* reported on 6 March 1894 that the **Newfoundland** had passed the Narrows going north and that she had "an advantage hardly fair to Newfoundlanders" because Newfoundland ships could not leave port before March 10. During the 1890s the **Newfoundland** cleared from Newfoundland ports or from Halifax. In 1903 it was reported that the **Newfoundland** would fit-out for the seal fishery in North Sydney, where she would secure a crew from the large number of Newfoundlanders employed in the different collieries there. Captain Farquhar was also involved in running the blockade to Cuba during the Spanish-American War in 1898 and while doing so the **Newfoundland** was caught by the Americans and taken into Charleston, North Carolina, where she was held for six months. In 1904 she was purchased by Harvey & Co. and had a moderately successful career as a St. John's sealer under the command of Captains Samuel Blandford, Job Knee and John Parsons.

In 1914, under the command of Captain Westbury Kean, the **Newfoundland** found herself jammed in the ice some distance from the other ships, which were obviously among seals. Captain Kean sent his men, under the command of second hand George Tuff, in the direction of the SS **Stephano**, which was under the command of his father, Abram. Captain Abram took them aboard and, in spite of the unsettled nature of the weather, carried them in the direction of a small patch of seals and put them on the ice. The men found themselves in a snowstorm and, unable to get to any of the ships, spent the following fifty-three hours on the ice. Seventy-eight men died (including one sealer who died after being rescued). The sight of the sealers' bodies being brought into port by the SS **Bellaventure** made a lasting impression on the inhabitants of St. John's and the **Newfoundland** disaster was extensively reported in the local press. An enquiry was held with inconclusive results.

In 1915 the **Newfoundland** was sold to Job Brothers & Co. Ltd. and her name was changed in 1916 to the SS **Samuel Blandford** in honour of that famous captain. On 4 August 1916 she was lost on the rocks in St. Mary's Bay with no loss of life.

32

The SS **Hope**, 307.35 net tons, 70 nhp, was a Peterhead whaler built in 1873 by A. Hall & Co., Aberdeen. She was 150 feet long, 28 feet wide and 17 feet deep. The **Hope** engaged in whaling and Arctic exploration, and on one of her trips to the Davis Strait had as her surgeon Dr. Arthur Conan Doyle, the creator of Sherlock Holmes. In 1892 the **Hope** was sold to Baine Johnston & Co. Ltd., St. John's, and for the remainder of her career she was employed as one of their "best, staunchest, and fastest" sealers, bringing in a total of 113,784 seal pelts. Among her sealing captains were William Bartlett Sr. and Abram Kean. In addition, she made several Arctic voyages as part of Robert Peary's expeditions and in 1894 brought the famous forty-ton meteorite from Cape York, Greenland, to New York City where it was placed in the Museum of Natural History.

In 1901, while at the ice-fields in the Gulf, the **Hope** was driven ashore on one of the Magdalen Islands and was a complete loss. Her crew suffered considerably, living in tents on Bryon Island while the storm raged around them, and their captain, William Bartlett, almost lost his life when he fell into an abandoned well on a neighbouring island. Captain Bartlett and most of his crew were brought to St. John's by the SS **Greenland**.

In this James Vey photograph, the **Hope** is on dry dock in St. John's, circa 1895.

33

The SS **Southern Cross**, 276.72 net tons (later reclassified at 325), 100 nhp, was built in Norway as a whaler and purchased for the G.E. Borchgrevink's British Antarctic expedition of 1898-1900. Acquired by Baine Johnston & Co. Ltd. of St. John's in 1901, she was sent to the seal fishery annually, bringing in a total of 132,060 seal pelts. Among her captains were Darius Blandford, Daniel Bragg and Peter Carter. Returning from the Gulf with a full load of seal pelts in 1914, the **Southern Cross** was lost in a storm with Captain George Clarke and 172 men. This was Newfoundland's worst sealing disaster.

34

The SS **Eagle II**, 418 net tons (reclassified at 457.77 in 1908 when her bow was extended), 82 nhp, was built in 1902 as the SS **Sophie** in Sandefjord, Norway. She was 176.2 feet long, 29 feet 8 inches wide, and 15 feet 1 inch deep. The **Sophie** was purchased by Bowring Brothers in 1903 and renamed the **Eagle**. Under this name she prosecuted the seal fishery for forty-four springs and brought in over 660,000 seal pelts, the fourth-largest number brought in by any steamer. Among her captains were Arthur Jackman, Albert Blackwood, Westbury Kean, Sidney Hill and Charles Kean.

In 1924 the **Eagle** carried the first aeroplane to the seal fishery, a *Gypsy Moth* model piloted by Roy Grandy of Grand Bank. It took two days to prepare a runway on the ice, but the plane found a patch of seals within six or seven minutes, returned to the runway and was hoisted back onto the **Eagle**. The ship's men killed and panned a full load of whitecoats and the steamer was first into St. John's carrying one of her heaviest loads of seal pelts. The **Eagle** was the last of the old wooden-wall steamers to remain in operation in Newfoundland and in 1950 Bowring Brothers, her owners, had her towed outside St. John's harbour and sunk.

In this March 1934 photograph, the **Eagle**, under the command of Captain Sidney Hill, is jammed in ice.

35

The St. John's fleet of wooden walls getting ready to leave port circa 1890, with one ship already through the Narrows. Note St. George's Hospital on Signal Hill. Photograph by James Vey.

36

The SS **Adventure**, 825.94 net tons, 213 nhp — the first iron-clad steamer to prosecute the seal fishery — was built by the Dundee Shipbuilder's Co. Ltd. in 1905 for Harvey & Co. Sent to the ice by Harvey's in 1906, the **Adventure** was 278 feet long, 38 feet wide and 25 feet deep, by far the largest sealing ship during the 1906-08 seasons. She engaged in the seal fishery for ten springs under the command of Captains Henry Dawe and Jacob Kean, bringing in a total of 181,578 seal pelts. She was sold to Russia in 1916.

Note the sidesticks which allowed the sealers to leave and board the ship rapidly.

37

The SS **Florizel**, 1980.22 net tons, 440 nhp, was a steel steamer built in 1909 by Charles Connell & Co., Glasgow and owned by the New York, Newfoundland & Halifax Steamship Co. Ltd., Liverpool. Known as the Red Cross Line (the red cross being the Bowring house flag), the company had been formed in 1884 with many Newfoundland shareholders and was managed by Messrs. C.T. Bowring & Co. Ltd. of Liverpool. The **Florizel** operated in the freight and passenger trade between St. John's, Halifax and New York and in the off-season was sent to the seal fishery by Bowring Brothers, St. John's.

During five of the eight years that the **Florizel** went to the ice between 1909 and 1916 she was by far the largest sealer. (The **Stephano**, at 2143 net tons, prosecuted the seal fishery in 1912, 1913 and 1914.) Under the command of Captain Abram Kean, the **Florizel** was the first ship to use wireless telegraphy at the seal fishery: on 18 March 1909, when the ship was 260 miles from Cape Race, Captain Kean sent the following message to Bowring Brothers:

> 17,000 panned, 12,000 on board; **Eagle** 20,000 panned; rest of your fleet in seals, have not done much; **Adventure**, **Erik**, **Iceland**, **Bloodhound**, **Vanguard** and **Diana** here; have no report; **Florizel** great success. Sgd. A. Kean.

During her sealing career she was commanded by two captains, Abram Kean and his son Joseph. In 1910 she brought in 49,069 pelts and in 1916 a further 46,481 (adding considerably to Abram's reputation), and had a total catch of 200,862 seal pelts.

At the beginning of World War I the first draft of the Newfoundland Regiment (later designated Royal), the Blue Puttees, sailed for England on the **Florizel**, 4 October 1914. In 1918, while en route to Halifax and New York, the **Florizel** went aground near Cappahayden, Newfoundland with the loss of ninety-four lives, including Captain Joseph Kean who was on board as a passenger.

38

The SS **Bellaventure**, 466.98 net tons, 356 nhp, was another of Harvey & Co.'s three steel ships. Built by D. & W. Henderson, Glasgow, she was 253 feet long, 36 feet wide, and 19 feet deep. Beginning in 1909 she made seven trips to the seal fishery and brought in a total of 112,135 seal pelts. In 1916 she was sold to Russia.

In this photograph the **Bellaventure** is entering St. John's in 1914, carrying the bodies of the seventy-seven sealers who had lost their lives in the **Newfoundland** disaster, as well as some of the survivors.

39

The SS **Bonaventure**, 461.04 net tons, 350 nhp, was built for Harvey & Co.'s Newfoundland seal fishery by Messrs. Napier, Miller & Co., Glasgow. Built of steel and heavily reinforced, she made her first trip to the ice-fields in 1909 and annually thereafter until the 1915 sealing season. In 1915 the **Bonaventure** was *highliner*, bringing in over 25,000 seal pelts while the other twelve steamers managed less than 12,000 among them. Her captain during the 1909-14 seasons was John Parsons and in 1915 she was commanded by Bob Bartlett. On 13 March 1913 the **Bonaventure** collided with the SS **Beothic I** in the St. John's Narrows while leaving for the ice-fields. While it is difficult to ascertain exactly what happened, it seems that the **Beothic** slowed and the **Bonaventure** was going too fast to stop. The **Bonaventure** suffered only slight damage and continued on her way but the **Beothic**, under the command of Captain W.C. Winsor, was seriously damaged. Forced to return, she went on dry dock for repairs and thereby missed the season's voyage. The **Bonaventure** was sold to Russia in 1916.

This photograph by Ern Maunder shows the two steamers at the time of the accident.

40

The SS **Sagona**, 420.4 net tons, 98 nhp, was a steel steamer built in 1912 for Crosbie & Co. by the Dundee Ship Building Co. She was especially designed for operating in ice and had the latest in passenger accommodations. She was sent to the ice by Crosbie & Co. between 1912 and 1914. In 1921 and 1922 she was sent to the ice by the Reid-Newfoundland Co. A number of different suppliers used the **Sagona** at the seal fishery until finally in 1938, supplied by E.S. Ullmann, she made her last trip. Under the command of a number of different captains, the **Sagona** brought in a total of 165,599 seal pelts.

Sealing made up only a small part of the **Sagona**'s career. In 1915 she was chartered for the Cabot Strait passenger run by the Reid-Newfoundland Co. She also served on the Newfoundland-Labrador service. In 1922 the **Sagona** rescued the men from the **Diana** (above photograph) and in 1931 was sent to assist at the **Viking** disaster. The **Sagona** was sold in the 1940s.

In this photograph by S.R. Oakley, sealers from the **Diana** can be seen moving their sealing boxes and clothes bags aboard the **Sagona**.

41

The SS **Nascopie**, 1004.01 net tons (reclassified to 1488 in 1928), 340 nhp, was built at Newcastle-on-Tyne during 1911-12 for the Hudson's Bay Co. She was of steel construction and was designed to ride up on the ice-floes and use her weight to break a track through the ice. One of the largest ships to prosecute the seal fishery, she was sent to the ice for four springs by Job Brothers & Co. Ltd. (1912-15); Bowring Brothers chartered her for the ice in 1927 and 1928; and Job Brothers & Co. Ltd. chartered her again in 1929 and 1930. During her career in the Newfoundland seal fishery she was commanded by two sealing captains, George Barbour and Abram Kean, and brought in a total of 186,229 seal pelts.

During the 1930s and '40s the **Nascopie** carried out fur trade activities in the Arctic and was used by the Canadian government to patrol the eastern Arctic. The **Nascopie** was active in both world wars and during World War I, armed with one gun, she sank a German submarine while in the White Sea. In 1947 the **Nascopie** was lost in a storm off Cape Dorset, Baffin Island.

Note the sidesticks in the above photograph.

42

The SS **Beothic II**, 1248 net tons (reclassified to 1078 in 1929), 150 nhp, formerly the **Lake Como**, was a steel vessel built in 1918 by the American Steamboat Co. of Lorain, Ohio. She was purchased for the seal fishery by Job Brothers & Co. Ltd in 1926 and considered a major addition to the Newfoundland sealing fleet, bringing in over 48,000 seal pelts on two separate occasions. Over a period of fifteen consecutive years, 1926-40, she brought in a total of 389,664 seal pelts. Her captains were George Barbour, Peter Carter, W.C. Winsor, Jacob Kean, Abram Kean, Sidney Hill and Stanley Barbour. In 1940, while under charter to the Canadian National Railway and commanded by Captain Stanley Barbour, the **Beothic** went aground at Griquet and was a complete loss.

In this photograph by S.R. Oakley, the **Beothic** is jammed in ice.

43

The SS **Caribou**, 1362 net tons, was built in Rotterdam, Holland, launched in 1925, and arrived in St. John's on 23 October of the same year. She was 265 feet long, 41 feet wide, and 27.5 feet deep. The **Caribou** served as a government coastal ship on the passenger service between Port aux Basques and North Sydney for almost sixteen years. In 1935 the Commission of Government chartered her to Crosbie & Co., who sent her on her only trip to the ice with Captain W.C. Winsor as master. She was highliner that year, securing 36,370 seal pelts, with the men's share being $56.99 each.

On 14 October 1942, while on the Cabot Strait run, the **Caribou** was torpedoed by a German U-boat and sank three to four minutes after the explosion. One hundred and thirty-seven lives were lost.

44

The SS **Imogene**, 715 net tons, 379 nhp, was built for Bowring Brothers by Smith's Stock Co., Southbank-on-Tees. She made her first trip to the seal fishery in 1929. While not as heavy as the larger sealers, this steamer was sufficiently powered and uniquely designed to operate in Arctic ice. In 1933 the **Imogene** brought into St. John's a record number of seal pelts — 55,636. During twelve trips from 1929 to 1940, the **Imogene**, under the command of three captains — Al Blackwood, Westbury Kean and Sidney Hill — brought in a total of 425,720 seal pelts. On 22 September 1940 she ran around off Canso, Nova Scotia, and was a total loss.

This Leonard Winsor photograph shows the **Imogene**, under the command of Captain Albert Blackwood, returning from the 1934 seal fishery with over 52,000 pelts.

45

The SS **Kyle**, 548 net tons, was built in 1913 by Swan, Hunter & Co., Newcastle-on-Tyne, for the Reid-Newfoundland Co. She was 220 feet long and 32 feet wide. The **Kyle** was strongly constructed for steaming through ice, and succeeded the SS **Bruce** on the run between Port aux Basques and North Sydney. In 1926 the SS **Caribou** replaced the **Kyle** on the Cabot Strait run and the latter was assigned primarily to the Newfoundland-Labrador run where, during the following decade, she became a familiar sight to Newfoundland's fishermen who travelled annually to the Labrador coast. In addition, she was employed in maritime emergencies, including those involving the sealing fleet. For example, she was sent with supplies to the **Beothic** when that steamer was jammed in White Bay in 1933. The **Kyle** was bought by a Canadian company in 1958 and renamed the **Arctic Eagle**, then purchased by Earle Freighting Services of Carbonear in 1960, renamed the **Kyle** and was sent to the seal fishery under the command of Guy Earle. In 1965, while at the seal fishery, she struck an iceberg in the Straits of Belle Isle and was seriously damaged. Shortly after her return to port, the **Kyle** went aground in Harbour Grace, where she remains as the last representative of Newfoundland's coal-burning sealers.

46

The Auxiliary Schooner **Young Harp**, 60 net tons, 80 feet long, 24 feet wide, nine feet deep, was built in 1926. In addition to the wooden and iron-clad steamers, there were a number of craft listed variously as schooners, sailing vessels and motor vessels (primarily from the outports) prosecuting the seal fishery. For example, in 1930 the **Young Harp**, under the command of Captain George Norman and carrying twenty-nine men, brought in a total of 3,238 seal pelts for Ashbourne Ltd; the crew received $81.25 each. She was lost in ice in the Bay of Islands in 1956.

This photograph, circa 1930, shows the **Young Harp** at the ice.

47

In the 1870s the Dundee Seal & Whale Fishing Co. decided to open in St. John's in order to take advantage of the availability of Newfoundland men and seal resources. The company prospered for about fifteen years but the loss of the sealing steamers **Arctic**, **Narwhal** and **Resolute**, combined with the decline in Arctic whaling, forced the company to wind up its business and sell its remaining sealing ships to Newfoundland firms.

In this photograph, circa 1880, are the company's captains and others associated with this firm.

Back row (left to right): Captain Alexander Graham, born in Newfoundland, master of Newfoundland sealing steamers including the colony's first, the SS **Bloodhound**, in 1863; Captain McLennan, a Scotsman, master of the SS **Narwhal** during the 1870s; Captain Bannerman, a Scotsman, master of the SS **Aurora** during the 1870s; Captain Charles Dawe, Newfoundland-born master of sealing vessels (see caption 72); Hon. Moses Monroe, one of the founders of the Dundee Seal & Whale Fishing Co.; Mr. John Pye, company representative from Dundee; Mr. A.G. Smith, broker.

Front row (left to right): Captain John Green, manager of the Tug Company; Captain Charles Yule, a Scotsman, master of the SS **Esquimaux** and the SS **Resolute**; Captain William Adams Sr., a Scotsman, master of the SS **Arctic**; Mr. Francis Winston, editor of the *Morning Chronicle*.

48

Two young sealers with their *hauling ropes* and *gaffs*. Sealing required agility and experience and most sealers went to the ice for the first time at a relatively young age, generally in their teens.

49

Men and ships ready to leave for the ice. A postcard made from this photograph states that the governor of Newfoundland is visiting the fleet prior to its departure, circa 1900.

50

During the days of the sailing ships and the wooden walls, the men were often called upon to help tow the ship out of the harbour into the open sea. In this photograph the men of the SS **Southern Cross** are on a towline assisting their ship through the ice in St. John's harbour. Even with her sails unfurled and her engine engaged, the **Southern Cross** needed the crew's help. Photograph by Holloway.

51

Sometimes it was necessary to saw channels in the ice in order to allow the ship to pass through. Saws like the one in the top photograph were made for this purpose. The bottom photograph by John C. Loveridge shows men clearing a channel in Twillingate harbour.

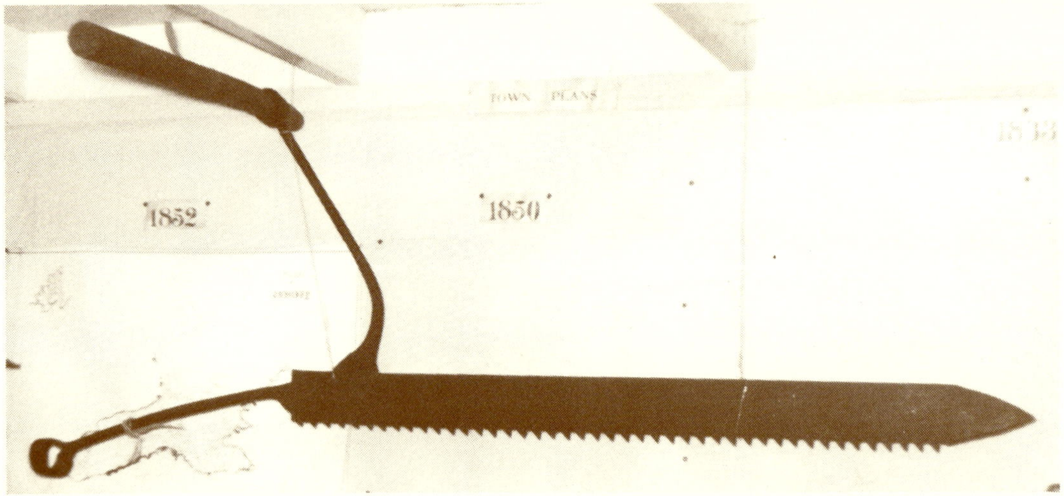

52

Whenever a sealing ship became jammed in the ice one of the standard methods used to free her was blasting the ice with gun powder. Deep holes were dug and cans of powder with long fuses attached were lashed onto sticks and pushed into these holes. In this 1934 photograph the men of the SS **Terra Nova**, under the command of Captain Stanley Barbour, prepare to blast the ship free. The man to the right, in the background, is holding a stick with a can of powder tied to it. Captain Carl Barbour of Newtown, Bonavista Bay, wearing the lighter suit and holding a pole, is standing with his back to Mr. Max Pardy of Musgrave Harbour who is smoking a pipe. Standing on deck near the rail, second from the right, is Mr. Arthur Andrews of Cape Freels, and on the ice, second from the left, is Mr. Baxter Vincent of Newtown.

53

The galley of a wooden wall — probably the SS **Vanguard** — on which W.H. Greene went to the ice in 1902. This photograph is from Greene's *The Wooden Walls Among the Ice Floes*.

54

Sealers waiting for the order to go overboard and begin their day's work.

55

Sealers with their gaffs *copying* over the ice in search of seals.

56

Sealers towing *laced-up* pelts. A *tow of pelts* generally consisted of two to four pelts depending on the size of the seals and the condition of the ice. Photograph by S.R. Oakley.

57

Seal pelts were usually piled on firm ice pans, hence the terms *pan of seals* or *pan of pelts*. Each pan was marked with the ship's flag which had the ship's initial, and the pelts would be picked up by the ship at the end of the day.

58

Single flags were used to mark pans of pelts and double flags were used to mark where the men would meet. In this photograph by Holloway Studios we see two flags of the SS **Grand Lake**.

59

Seal pelts were tied together and hoisted aboard by a steam-driven winch.

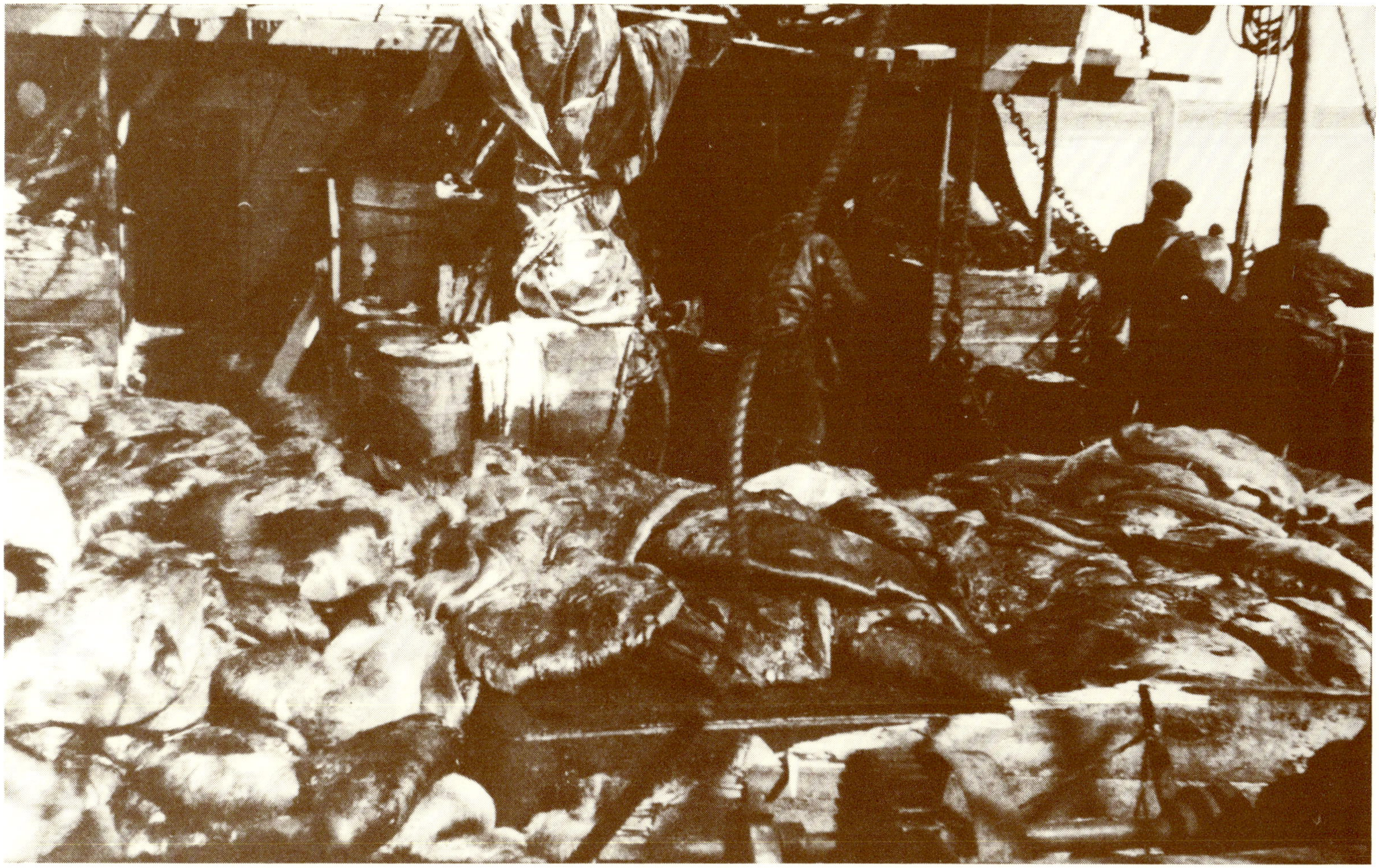

60

After the pelts were hoisted aboard the vessel, it was usually necessary to allow them to cool before stowing them in pounds below deck. Photograph by S.R. Oakley.

61

After the seal pelts had cooled off completely they were stowed in pounds below deck. The first layer was placed with the fat side down, but the remaining layers were placed fat side up. Chopped ice was shovelled over each layer of pelts. The space below deck was always divided into pounds so that the heavy pelts could not slide around and destabilize the ship.

Informed sources described the boards dividing the pounds in the **Southern Cross** as being in a weakened condition, and these sources suspected that, in the storm in which the sealer found herself, the boards gave way, the seal pelts slid sideways, and the ship capsized (see caption 33).

62

Collecting the ice which was spread over the pelts as they were stowed in the pounds was one of the sealers' regular duties. This 1960s photograph shows one crew gathering ice for this purpose. On the wooden walls, and on most of the iron-clad steamers, ice was gathered for cooking and drinking also. This ice was cut from high pinnacles from which most of the salt had drained, hence the sealers' term *pinnacle tea*.

63

While engaged at the seal fishery the men usually tried to procure a barrel of seal carcasses to bring back home. Each carcass was chopped up, stowed in the barrel, and preserved with either pickle from the galley or with ice.

64

If the ship's load was still incomplete after the season for whitecoats finished, gunners were sent on the ice to shoot the older seals. Each gunner was accompanied by two or three men, referred to as *dogs*, who carried the cartridges and who kept a count of the number of seals killed. The photograph on the right was taken circa 1930; the one below shows a muzzle-loading rifle of the type used by sealers in the nineteenth century.

65

Beginning in the 1830s, seal pelts were sold in Newfoundland by weight. Only in recent years have they been sold by number and type. In this photograph men are busy unloading and weighing seal pelts on a St. John's wharf. Note the scales, generally referred to as the *jig*.

After the pelts were landed, seal skinners were employed to remove the fat from the skin, usually with specially designed seal-skinning knives. Because they usually found employment at other trades near sealing centres (especially St. John's), seal skinners did not go to the ice but were hired on a temporary basis during the sealing season. A long apprenticeship was required and, by the 1920s, an apprentice could only be hired as a seal skinner if one of the older members died. Seal skinners could *defat* or *skin out* about 450 young harp pelts in a ten-hour work day. By the late 1920s machinery had replaced these craftsmen.

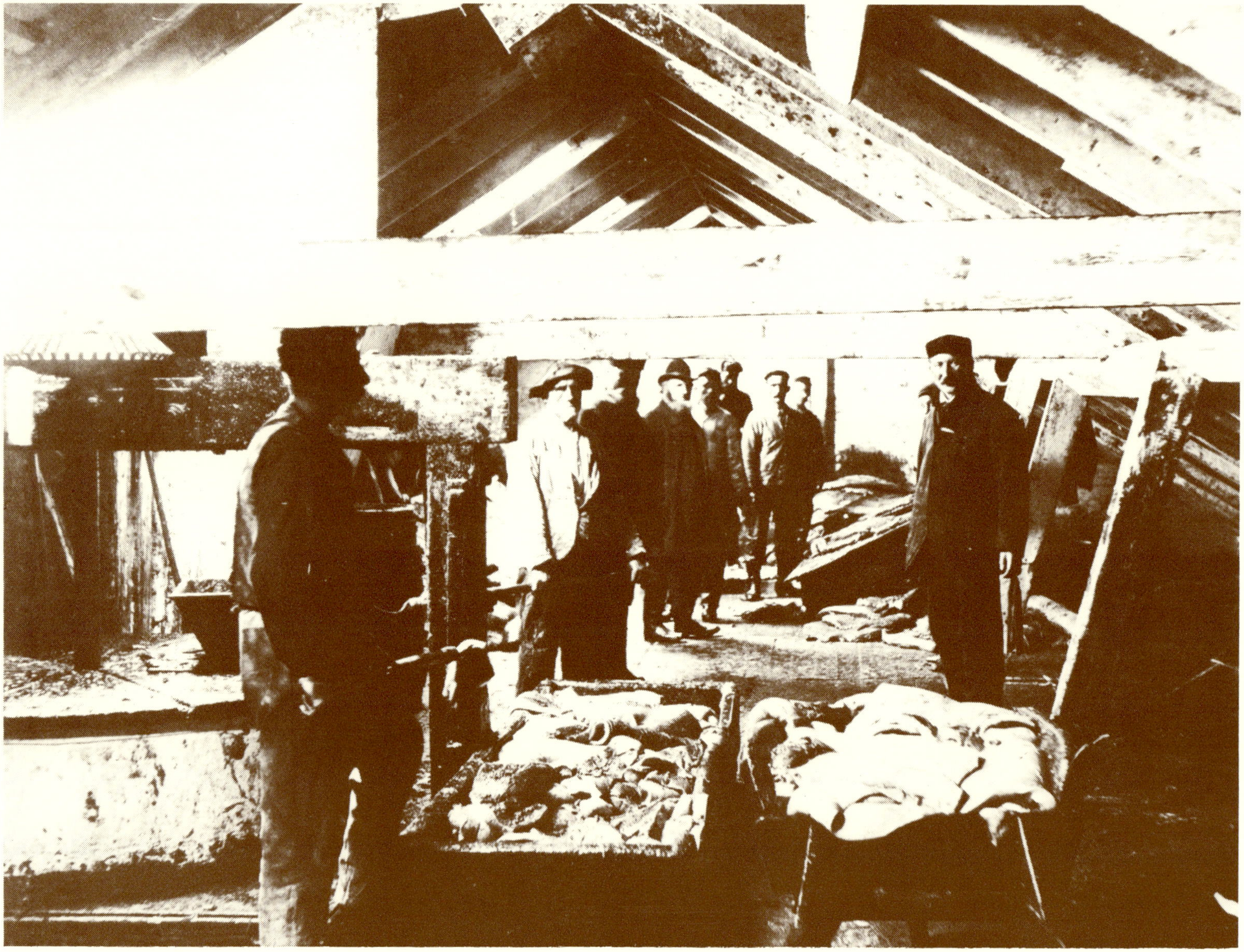

67

Munn & Co., Harbour Grace, was noted for producing some of the highest quality seal oil in Newfoundland. Even after the company's bankruptcy, the sealing premises continued to be used by Baine Johnston & Co. Ltd. This photograph, taken by James Vey circa 1900, shows the Harbour Grace seal-skinning loft.

68

Seal flippers are generally considered a delicacy in Newfoundland. Sealers always left one flipper attached to each pelt. When the pelts were unloaded in St. John's, these flippers were then cut out and shared among the sealers for their own use. While some were kept for personal consumption or given to relatives and friends, many were sold to local householders. Since this money was received in addition to their share from the voyage, some sealers, while still in St. John's, spent some or all of their *flipper money* on rum.

69

The most important dramatization of Newfoundland's traditional seal fishery is Varick Frissell's movie, *The Viking*. The idea for this movie apparently originated from conversations between Samuel Robert Oakley, a local photographer, and Frissell, a native of Massachusetts. Frissell first came to Newfoundland as a volunteer with the International Grenfell Association in Labrador in 1922. Over the next few years he captured some of his journeys on film, resulting in his 1926 release *The Lure of the Labrador*. In 1927 he and Oakley went to the seal fishery in the SS **Beothic** under the command of Captain George Barbour, Oakley's brother-in-law. Frissell secured about seven thousand feet of film covering all phases of the sealing industry, including the unloading and processing in St. John's. The final result, *The Great Arctic Seal Hunt*, is an excellent record of the seal fishery in general and of the 1927 season in particular. This trip gave Frissell an opportunity to get first-hand knowledge of the seal fishery and to acquire the experience needed for successful filming under these unique and difficult conditions. Contemporary reports suggest Frissell followed the suggestion of Oakley who encouraged him to produce a movie at the ice-fields and in Newfoundland. Frissell prepared a script and made arrangements to begin at the 1930 seal fishery.

Frissell arrived in St. John's on 16 January 1930, and was soon followed by his staff, actors and actresses. The principal performers included Charles Starrett, Louise Huntington,

Left to right: George Melford, Director; Maurice Kellerman, Cameraman; Varick Frissell; S.R. Oakley.

Movie scene with Signal Hill in background.

Production crew on the ice. Camera on the left, sound equipment on the right, and additional equipment under tarpaulin in foreground.

Filming beside the **Viking**. Most filming on the ice occurred near the ship.

Arthur Vinton, Edgar Nelson, and the famous Newfoundland explorer, Captain Bob Bartlett. Asked by a reporter if he was going to the ice as a sealing captain, Bartlett replied, "No. I came here to be a movie star — a real honest-to-goodness actor." Frissell worked until the end of February filming scenes in and around St. John's and Quidi Vidi, and then in March he and his party accompanied the fleet to the ice on board the SS **Ungava** (supplied by Job Brothers & Co. Ltd. and commanded by Captain W.C. Winsor).

This trip was a unique experience for all concerned. Captain Winsor and his sealers, anxious to fill the ship with seal pelts, were a little impatient with the film crew at times, but on the whole they found the filming an interesting diversion. Frissell rehearsed the actors and filmed his scenes, but discovered that his cast and crew had great difficulty copying on the ice. Consequently, sealers were called on to take some minor parts and although some refused, others joined in the activities with relish. One sealer who was present reports that the men laughed with glee when Captain Bartlett showed difficulty learning his lines. Frissell and his men were pleasant to work and live with, and all went well until the sealing was finished: since it was getting late in April, the sealers began agitating to return to St. John's.

Captain Winsor contacted Job Brothers & Co. Ltd. and it was decided to send the SS **Viking** to the ice to take the movie crew so they could continue filming while the **Ungava** returned to St. John's with the sealers. The film equipment, trunks and cases were put on the ice in preparation for the transfer. At this point, the sealers clamored to be paid for their extra work, trouble, and the additional time they had already spent at the ice. Captain Winsor informed Frissell that a signed agreement promising the men compensation was required before he would allow the equipment and other property to be put aboard the **Viking**. When some of the **Viking**'s crew tried to transfer the equipment to their ship before any agreement was signed Winsor stopped

69 (continued)

them, reportedly telling his crew, "Let it go to the Jesus bottom! You haven't got to work for the Yanks for nothing." When an agreement was signed promising each man $5 extra from Job Brothers and $5 from Frissell's party, the transfer was completed.

The **Ungava** returned to St. John's on April 21 with the largest number of seal pelts in the whole fleet — 37,177. In addition to the $5 bonus from their employer, Job Brothers & Co. Ltd., the men received $77.86 as their share of the catch. However, it appears that they may not have received the $5 bonus promised by Frissell: the **Viking** did not return to port until May 3 and by this time most of the **Ungava**'s men had returned home.

After leaving the **Ungava**, Frissell's party proceeded northward with Bartlett in command of the **Viking**. About seventy men had been hired in St. John's and when the ship arrived in St. Anthony more scenes were shot with these men and about fifty area residents. It was here that Starrett's character, Luke, climbed to the top of an iceberg to scan the horizon for seals. The **Viking** eventually returned to St. John's where a few final scenes were filmed.

The Viking was the first feature-length sound film made in what is now Canada and the first to use sound on location. Frissell's concern for authenticity and accuracy resulted in some remarkable film footage of the sealers' activities and of the ship's difficult passage through the ice-floes. Although the script is a typical melodrama of the kind produced in the 1920s, *The Viking* is a significant work — both for its contribution to the study of the history of the seal fishery and for its place in film history.

It seems certain that S.R. Oakley remained closely associated with Frissell during the filming. Rare Oakley photographs in the Cater Andrews' collection include pictures of Frissell and his party at Signal Hill, Quidi Vidi, and the ice-fields, as well as production stills from Quidi Vidi, the ice-fields and St. Anthony's harbour.

Filming of one of the storm scenes. On the right, artificial snow is being blown by a wind machine constructed from an aeroplane propeller.

Film crew's equipment and belongings on the ice prior to its transfer to the **Viking**.

The **Viking** with film crew in St. Anthony.

Oakley and Frissell appear in a group photograph, and another shows Oakley with the SS **Viking** in the background.

Upon Frissell's return to the United States he found that Paramount Pictures was not entirely satisfied and they advised him to shoot more action footage at the ice, including, if possible, an iceberg capsizing. As a result, Frissell returned in 1931 and went to the ice on the SS **Viking** under the command of Captain Abram Kean, Jr. Extra gun powder, for creating action scenes, was carried in addition to the ship's usual supply. While proceeding to the ice-fields, near the Horse Islands, the steamer blew up and sank. Twenty-four men, including Frissell, were killed. *The Viking* and *The Great Arctic Seal Hunt* remain unique records of the seal fishery in the days of the wooden-wall steamers. Oakley's photographs provide us with a further record of Frissell's work on sealing and the people involved in it.

S.R. Oakley and the SS **Viking**.

70

Captain Edward White, Sr. was born 9 October 1811 at Tickle Cove, Bonavista Bay, to William and Mary White. In 1835 he married Anne Weir, St. John's. They had eight children, including Edward Jr. and Richard, who also became captains.

In 1836 he went to the seal fishery for the first time as captain of a sailing vessel, and for the next forty-five years he served as a sealing master, first in sailing vessels and later in steamers. Between 1866 and 1882 (inclusive) Captain White brought in over three hundred thousand seal pelts in the steamers **Hawk**, **Nimrod**, **Hector**, and **Neptune** (the latter three constructed in Dundee under his personal supervision). Captain White, like other sealing masters, was occasionally caught in sudden storms, and this was the case in Green Bay in 1864 when his ship was lost along with twenty-five others. On that occasion, fifteen hundred shipwrecked men, including Captain White and the other sealing masters involved, were landed at Greenspond. Captain White also prosecuted the cod fishery from an early age and eventually moved to St. John's where he became a leading political and religious figure. He was appointed to the Legislative Council in 1861 and the Executive Council in 1882. In this capacity, he introduced the first sealing legislation which, among other things, fixed an opening date for the spring seal fishery.

Captain White died in St. John's on 1 June 1886.

71

Captain Azariah Munden was born 3 April 1813 at Brigus, Conception Bay, to Mr. and Mrs. William Munden. William, a well-known sealing master and entrepreneur, had the first Newfoundland sealing ship to exceed 100 net tons built in 1819. This ship, the **Four Brothers**, 104 net tons, was rebuilt by Azariah in 1851 and rechristened the **Three Sisters**.

Captain Munden was a progressive and astute sealing captain. Even before the adoption of steamers by the St. John's firms he advised Punton & Munn of Harbour Grace to acquire this type of vessel. The company had so much money invested in sailing ships that it was later — and then only gradually — that it and its successor, John Munn & Co., began to invest in steamers. When they purchased their first steamer they christened it the **Commodore** in Azariah's honour, and put him in command. In the **Commodore** his success as a sealing captain was greater than ever and in 1872 he brought in the greatest number of seal pelts ever for one trip into Harbour Grace — 31,314. His men made $158.21 each. The *Boston Traveller* described the event:

> ...The sealers crammed every nook and cranny of the **Commodore** with the precious fat. When the ship was filled below decks pounds were erected on deck; not content even with this a pile of some 1,400 seals, which remained, was made fast to the steamer and towed behind. The decks of the ship thus heavily loaded, were below the water, the bulwarks alone keeping out the wash of the sea.

The following year Captain Munden was sent to Aberdeen to supervise the construction of a new vessel and although John Munn told him to christen her the **Admiral**, Captain Munden decided to call her the **Vanguard** instead. He commanded this steamer until 1879 when he retired.

He died 4 August 1889.

72

Captain Charles Dawe was born 28 February 1845 at Port de Grave, Conception Bay. He was involved in sealing and fishing from an early age and commanded the sailing vessels **Huntsman** and **Rolling Wave** at the seal fishery. In 1871 he married Miss Emma Bartlett; they had one son and two daughters.

In 1874 he was placed in charge of a sealing steamer for the first time. A popular and successful captain, over a twenty-two year period he commanded the sealing steamers **Greenland**, **Aurora**, **Iceland**, **Thetis**, **Bear**, **Terra Nova**, **Newfoundland** and **Vanguard** and brought in a total of 318,280 seal pelts for Harbour Grace and St. John's firms.

Besides his activities in the seal fishery, Captain Dawe, through his firm C.A. Dawe of Bay Roberts, also built up a sizeable business in the Labrador cod fishery. A successful politician, Dawe served as the member for Harbour Grace from 1878 to 1889 and later represented Port de Grave. In 1893 he was appointed to the Executive Council and later became Leader of the Opposition.

He died 30 March 1908.

73

Captain Samuel Blandford was born 10 August 1840 at Greenspond, Bonavista Bay, and later married Miss S.A. Edgar of that community.

He began his sealing career at the age of thirteen when he went to the ice with his father in their schooner the **Vulcan**. He continued to participate in the sailing ship seal fishery and served on a number of different vessels. In 1873 he served as second hand with Captain Isaac Bartlett on the SS **Tigress** and in 1874 he went master of the SS **Ospray** (or **Osprey**), which he lost at the ice that same year.

Captain Blandford went in command of the SS **Iceland** in 1876, and until 1906 (thirty-one years) he commanded sealing steamers, mainly for Job Brothers & Co. Ltd. He captained the SS **Eagle** in 1879 and the SS **Esquimaux** from 1880 to 1882. Captain Blandford succeeded Captain Edward White Sr. on the SS **Neptune** in 1883 and commanded her annually until 1903. In 1904 he served as master of the sealing steamer **Newfoundland** and the following two sealing seasons, 1905 and 1906, he commanded the SS **Virginia Lake**.

During his career in command of sealing steamers Captain Blandford brought in a total of 604,775 seal pelts — the fourth-highest number brought in by any steamer captain. His best years were 1884 when he brought in 41,983 (his men each received $110.55) and 1888 when he returned with 42,242 (the men's share was $77.17). He retired in 1906.

Throughout his sealing career Captain Blandford worked as manager of Job Brothers & Co. Ltd. summer fishery in Blanc Sablon. He was elected to the House of Assembly in 1889 and appointed to the Legislative Council in 1893.

Captain Blandford died 8 March 1909 and in 1916 the SS **Newfoundland** was rechristened the SS **Samuel Blandford** in memory of this great captain.

74

Captain Arthur Jackman was born in 1843 to Captain and Mrs. Thomas Jackman of Renews. He was introduced to the seal fishery at an early age and was only twenty-two years old when he commanded his first schooner at the ice.

"Viking Arthur" was appointed captain of the sealing steamer **Hawk** in 1871 and went on to command the steamers **Falcon**, **Narwhal**, **Resolute**, **Eagle**, **Aurora**, **Terra Nova**, and **Eagle II**. From 1871 to 1906, thirty-six years, Captain Jackman never missed a spring at the ice, bringing in 533,190 seal pelts.

His career was not uneventful. In 1881, while coming from Scotland in the SS **Resolute**, Jackman ran into a storm during which a heavy sea washed away the ship's funnel. He immediately had his men construct a makeshift funnel out of packing cases lined with sheet iron and tin cans and continued on to Newfoundland. He later lost the **Resolute** at the seal fishery in 1886 and the SS **Eagle I** at the whale fishery in 1892. More notably, he commanded Commodore Peary's ship on Peary's first polar venture in 1886. In 1906 he was fined $2,000 for killing seals on a Sunday in violation of sealing laws.

Captain Jackman's life was tied to the sea and he commanded different Newfoundland coastal steamers, including the **Curlew** and the **Plover**, and served as marine superintendent for Bowring Brothers.

He died 31 January 1907.

75

Captain George Barbour was born at Cobbler's Island, Bonavista Bay, 27 July 1858 to Benjamin and Rebecca Barbour. He married Miss Lucretia Jane Oakley, the sister of photographer Samuel Robert Oakley. The couple had three sons and one daughter.

Job Brothers & Co. Ltd. appointed him captain of the SS **Walrus** in 1893 and he went on to spend thirty-seven springs at the seal fishery, bringing in a total of 752,563 seal pelts — second only to the record eventually set by Captain Abram Kean. His largest catch was 48,421 pelts, brought into St. John's aboard the SS **Beothic II** in 1926, earning his men $124.66 each. He also commanded the SS **Neptune** in 1925, the year she brought in the one millionth pelt of her sealing career. In 1927 Varick Frissell accompanied him to the seal fishery (see caption 69). Captain Barbour's most tragic experience at the seal fishery occurred in 1898 when he lost forty-eight men in the **Greenland** disaster. Over the course of his career as a successful sealing master, Captain Barbour commanded the sealing steamers **Walrus**, **Greenland**, **Vanguard**, **Bloodhound**, **Neptune**, **Beothic**, **Nascopie** and **Beothic II**. He also prosecuted the cod fishery and was employed in the northern Labrador coastal service.

Captain Barbour died at Trinity, Trinity Bay, in 1928.

76

Captain Abram Kean was born 8 July 1855 at Flower's Island, Bonavista Bay to Joseph and Jane Kean. The youngest of nine children in a family of cod and seal fishermen, he entered the fisheries with his brothers at an early age. Abram was unique, however, in that he was the only one of the family to receive a regular classroom education. At the age of seventeen he married his father's housekeeper, Miss Caroline Yetman, who was seven years his senior; they had eight children — six sons and two daughters.

After being involved in the fisheries with his brothers and having served as master watch, second hand, and captain of sailing vessels in the seal fishery, Abram applied for a position on a steamer and in 1884 went as bridgemaster with Captain Joe Barbour. He began a political career with his election to the House of Assembly as member for Bonavista in 1886, as a result of which he moved to St. John's. While there he also qualified for his Master Certificate of Competency in 1888.

In 1889 he was appointed captain of the SS **Wolf** and after a record-setting round trip lasting only eleven days he arrived in St. John's on 20 March with 26,912 seal pelts. During a career as sealing captain which lasted over forty-eight consecutive springs, ending in 1936, Captain Kean commanded nine steamers: the **Wolf**, **Hope**, **Aurora**, **Terra Nova**, **Florizel**, **Stephano**, **Nascopie**, **Thetis** and **Beothic II**. The total number of seal pelts he brought in — 1,052,757 — is still the largest number by any sealing captain. In 1910 Captain Kean brought in 49,069 pelts in one trip in the **Florizel**. This remained a record until Captain Al Blackwood in the SS **Imogene** and Captain Peter Carter in the SS **Ungava** both broke it in 1933. Captain Kean brought in over 48,000 seal pelts in 1934 as well.

In 1914 his part in the **Newfoundland** disaster (see caption 31) hurt his reputation and has made him a highly controversial figure. His position as "Commodore of the Fleet," however, became recognized and accepted.

Captain Kean worked at the cod fishery for about twenty years, the mercantile business for fourteen years, and spent nineteen years in charge of Newfoundland's coastal service. Besides representing Bonavista in the House of Assembly from 1885 to 1889, he represented Bay de Verde from 1897 to 1900. During this latter term he was acting minister of the newly-created Department of Marine and Fisheries and, as such, helped prepare the sealing legislation which was passed in 1898. In 1927 he was appointed to the Legislative Council. Captain Kean had the rewarding experience of seeing three of his sons — Joseph, Nathan and Westbury — become prominent sealing captains as well. He published his autobiography, *Old and Young Ahead*, in 1935 at the age of seventy-nine, and retired from service as an active sealing captain the following year. He died in 1945.

77

Captain Robert A. (Bob) Bartlett (above left) was born 15 August 1875 in Brigus, Conception Bay, to Captain William and Mary (nee Leaman) Bartlett, a family that lived from the proceeds of the seal and Labrador cod fisheries. After completing his early education, he embarked upon a career as fisherman, sealer, sailor and mariner. It is as an Arctic mariner and explorer that he is best known.

After he received his master's certificate in 1898 he served as first mate on the SS **Windward** during Robert E. Peary's 1898-1902 Polar expedition. During 1905-06 he commanded the SS **Roosevelt** on another major Polar expedition with Peary, and in 1908 they returned to the Arctic for their final attempt to reach the North Pole. Captain Bartlett brought the **Roosevelt** to Cape Sheridan, arriving on September 5. From there they moved overland to Cape Columbia, ninety miles away. In February 1909 Bartlett began the work of pioneering the trail to the Pole with Peary following behind. When they were less than 150 miles from the Pole, Peary sent Bartlett back and continued on with a small party. He claimed later that he reached the Pole around April 6. By the time this expedition was over, Bartlett was internationally recognized as an Arctic mariner. In 1913 he commanded the SS **Karluk** on an expedition to the western Arctic under Vilhjalmur Stefansson. The **Karluk** was crushed in ice on 10 January 1914 and, after an incredible journey over the ice Bartlett reached Alaska on May 28; he was able to rescue the few survivors from the expedition. Shortly afterwards he became an American citizen and made New York his headquarters. During the 1920s and 1930s he continued to explore the Arctic, sponsored by different groups (including the American Museum of Natural History, the University of Michigan, the Museum of the American Indian, the American Geographical Society, the Smithsonian Institute, the New York Botanical Gardens, the Chicago Zoological Society and the Philadelphia Academy of Natural Sciences). He made twenty-eight voyages to the Arctic, twenty-two of them within Canadian territory, and received a number of prestigious awards and medals for his work.

As a young man Bartlett had gone to the seal fishery and eventually became a sealing captain, but his career as captain at the ice was confined to seven trips between 1901 and 1915, during which time he brought in a total of 80,093 seal pelts. In 1930 he returned to Newfoundland to play the part of Captain Barker in Frissell's movie *The Viking*.

Bartlett continued to live in New York when he was not exploring. He wrote his autobiography, *The Log of Bob Bartlett*, lectured widely, and wrote an article on the Newfoundland seal fishery for *National Geographic*. He never married, and died in New York on 28 April 1946 from pneumonia.

78

Captain William C. Winsor's house, Wesleyville. Homes of prominent sealing captains were often major landmarks in their home communities.

Captain Winsor (inset) was born into a sealing/fishing family in Wesleyville in 1876. He married Miss Josephine Blandford, daughter of Samuel Blandford, in 1902; they had two sons and four daughters. He commanded his first sealing vessel, the **Virginia Lake**, in 1903 and then went on to command many of Newfoundland's major sealing steamers, including the **Bloodhound**, **Greenland**, **Beothic**, **Thetis**, **Neptune** and **Ungava**. In 1930 Captain Winsor had on board the **Ungava** Varick Frissell and Captain Bob Bartlett, who were engaged in the production of *The Viking*. In 1935 he commanded the **Caribou** on that ship's only voyage to the seal fishery. He was a highly successful captain and in 1938 brought in 47,534 seal pelts in the **Ungava**. He continued to command sealing ships up to the end of the 1949 season, when he was master of the **Arctic Prowler**. Captain Billy Winsor was also a prominent Newfoundland politician. He was elected to represent Bonavista in 1908 and again in 1909 but was defeated by Sir William Coaker in three subsequent elections. Re-elected in 1924, he became minister of Marine and Fisheries in Walter Monroe's government. He was defeated in the 1928 election but elected for the last time in 1932. After Newfoundland adopted Commission of Government in 1934 he returned to the seal fishery and commanded vessels until his retirement at the age of seventy-seven.

He died in 1964 at eighty-seven years of age.

79

Captain Albert Blackwood was born in Brookfield, Bonavista Bay, in 1884. He became a fisherman/sealer and in 1926 took command of the SS **Eagle**. In 1929 he took the SS **Imogene** to the ice and was her captain every spring (excluding 1931) until 1937. In 1933 Captain Blackwood brought in the record number of seal pelts — 55,636 — and was highliner again the following year with 52,030. His total catch of 366,096 pelts during eleven springs to the ice established a record annual average of 33,281 seal pelts.

He died in 1938.

80

Captain Peter Carter was born on Newell's Island, Bonavista Bay in 1869. He first went sealing in the SS **Commodore** in 1882 and was given command of the SS **Southern Cross** in 1907. During 1925-28 he was in charge of the SS **Thetis**, then spent three seasons on the SS **Beothic** (1929-31), and finally three on the SS **Ungava** (1933-35). In 1933, while employed by Crosbie & Co., Captain Carter secured the heaviest load of seal pelts ever brought in to a Newfoundland port. En route to St. John's from the Front with a cargo of seal pelts, he came upon a herd of older seals on tight ice and his men killed a considerable number. His final cargo contained 26,100 young harps, 4,176 old harps, 18,840 bedlamers, 138 young hoods, and 31 old hoods for a total of 49,285 seal pelts. His men received $80.36 each. During his eleven years as captain, Peter Carter brought in 239,459 seal pelts. In addition to his relatively short career at the ice, he prosecuted the Labrador cod fishery on a regular basis.

Captain Carter died in 1959.

The SS **Grand Lake** was damaged by the ice and sank on 4 April 1908 with 19,600 seal pelts on board. Her 203 men escaped to the SS **Vanguard**. Photograph by George Harding.

The SS **Iceland** was crushed by ice on 30 March 1910 and sank while burning with 4,500 seal pelts on board. Her captain and 137 men walked about two miles to the SS **Eagle**.

Unidentified photograph showing the crew of a sealing ship on the ice. The presence of their clothes bags and sealing boxes indicates that they have abandoned ship and are waiting to be picked up by another ship.

The SS **Eagle** was towed outside St. John's harbour and sunk on 23 July 1950 because it was no longer profitable to maintain her.

81

Newfoundland sealing ships and crews prosecuted a difficult and dangerous fishery and losses were high. In 1845, for example, the barque **Ringwood** was lost with all hands, the brig **Peerless** was lost with her captain and forty men, and the brig **Elizabeth Margaret** was lost with her captain and fifteen men. The steamers were not spared either: some were squeezed by the tightening ice and damaged so badly that they sank when the ice slackened; others were caught in running ice and swept along into dangerous shoals; propeller blades and rudders were lost; fires and gunpowder explosions were constant threats. Each of these photographs shows the end of a sealing ship and/or a related incident.

The MV **North Star VI**

The MV **Sir John Crosbie**

The 1940s, especially the period of World War II, was a watershed in the recent history of the Newfoundland seal fishery. The old coal-burning steamers had become too expensive and inefficient and most could no longer be employed profitably in any activity other than sealing. Therefore, as they were lost to the seal fishery they were not replaced and the last one, the SS **Eagle II** was sunk in 1950 by her owners, Bowring Brothers.

The sealing steamers were replaced by motor vessels. These were clean, efficient, and capable of meeting Newfoundland's growing demands for coastal freight and passenger transport. Also, the war pulled Newfoundland out of the Great Depression; there was a general growth in the economy which encouraged new investment in motor vessels. Some of this investment came from the outports and, as a result, St. John's was no longer the only centre for the spring seal fishery. Like the situation in the early nineteenth century, smaller outport entrepreneurs were encouraged to invest in smaller ships that could be used for many purposes, sealing being just one.

In the meantime, the sealing industry continued to decline in real terms, even though Newfoundland's total economy was growing. World War II brought prosperity to Newfoundland as Canadians and particularly Americans built military bases in the colony and offered employment to practically every able-bodied man and woman who applied. Fishermen deserted the fishing grounds, abandoned their boats and equipment, and became wage earners at bases such as Argentia, Placentia Bay; Fort Pepperrell, St. John's; and Harmon, Stephenville. Even after the war ended, the bases continued to function and employ thousands of Newfoundlanders. This declining interest in the fisheries generally was accompanied by a continued decline in the value of the

The MV **Chesley A. Crosbie**

The MV **Arctic Prowler**

The MV **Lady Johnson II**

The MV **Algerine**

sealing industry in particular. Nevertheless, it survived, thanks to a new generation of captains and ships. Captains and ships that were employed at sea in other capacities during ten months of the year were available to go to the ice in the spring, and although the number of sealers was declining there was always sufficient manpower to meet this limited demand.

As St. John's was losing control over Newfoundland's seal fishery, Newfoundland itself was losing control over what had always been its own seal fishery. By the late 1940s sealing ships sailed from the capital, from various outports, and from Canada's mainland. However, these Canadian ships depended upon Newfoundland captains and sealers. For example, in 1948, out of nine departures for the ice-fields reported in the local press, three of the ships were owned in Quebec, three in St. John's, and one each in Port Union, Catalina and Twillingate. Of the three from St. John's, two were sent out by Bowring Brothers and one by Crosbie & Co. These ships varied in size from the small MV **Catalina Trader** with twenty-nine men to the SS **North Voyageur** with 180. The nine captains and their ships were as follows: Leslie Winsor (MV **Linda May**); Jabez Winsor (MV **Lady Cecil**); James Gillett (MV **Bessie Marie**); Stanley Barbour (SS **North Voyageur**); George Norman (MV **Catalina Trader**); Llewellyn Kean (SS **Magdalen**); Wilfred Barbour (MV **Terra Nova**); J.H. Blackmore (MV **J.H. Blackmore**); and Ralph Smith (MV **Lady MacDonald**).

In 1957 there were no local reports of Quebec ships participating in the seal fishery out of Newfoundland ports, but Halifax had a larger fleet than all Newfoundland ports combined. In that year, five Newfoundland motor vessels cleared from St. John's: **Bessie Marie** (Captain Wilfred Barbour); **Terra Nova II** (Captain Gus Carter); **Algerine** (Captain Harold Laite); **Placentia** (Captain Abe Kelloway); and **Glenwood** (Captain Max Frampton). The **Arctic Prowler**, a Halifax ship under the command of James Gillett and carrying a crew of Newfoundland men, also cleared from St. John's; ten other sealing vessels cleared from Halifax. In the meantime, the Norwegians, whose participation in the seal fishery off Newfoundland's coasts had varied since earlier in the century, were now, in the 1950s, expanding their activities and some of their ships were operating out of Halifax.

Halifax became a sealing centre of considerable importance during the 1950s with both Canadian and Norwegian firms becoming established there. Newfoundland sealers continued to find berths on these ships as did a number of Newfoundland captains. Those Newfoundlanders who preferred to go sealing in the Gulf would go to Halifax and sign on the ships that were leaving for that area. At the end of the season they would be dropped off at Port aux Basques or St. John's. For example, Captain James Gillett of Twillingate and his sealing crew were employed regularly by Halifax firms and Captain William Moss of Princeton, Bonavista Bay, commanded the MV **North Star VI** for a number of years out of Halifax. Other Newfoundlanders preferred to

Captain Leslie Winsor (1906-1986). Born Wesleyville, Bonavista Bay, Newfoundland. Sealer and Labrador cod fisherman. Captain of the MV **Linda May** and the MV **Sir John Crosbie**.

Captain James Gillett (1909-). Born Twillingate, Notre Dame Bay, Newfoundland. Seaman. Captain of the MV **Arctic Endeavour**, the MV **Bessie Marie**, and the MV **Arctic Prowler**. Retired.

Captain William Moss (1911-1969). Born Princeton, Bonavista Bay, Newfoundland. Captain of the MV **North Star VI**, the MV **Arctic Sealer**, and the MV **Arctic Prowler**.

82 (continued)

wait in St. John's and sign on those Halifax sealing ships that were proceeding to the Front.

During the 1960s Newfoundland's fleet of sealing ships declined. Captain J.H. Blackmore gave up in 1962 and this left only Bowring Brothers and Earle Freighting Services. Crosbie and Co. sent out the **Sir John Crosbie** in 1963 and by the end of the decade only Crosbie & Co. continued to prosecute the seal fishery, sending out one vessel — the impressive **Chesley A. Crosbie**. In 1969 this vessel, under the command of Captain Clarence Dyke, prosecuted the seal fishery with eighty sealers on her first trip and forty sealers on her second. Meanwhile, Captains Gillett and Moss, among others, were continuing to work for Halifax firms.

In 1970 the MV **Lady Johnson**, owned by Captain Clayton M. Johnson and commanded by Captain Morrissey Johnson, joined the **Chesley A. Crosbie** at the ice. Captain Morrissey Johnson continued to take the **Lady Johnson**, and after she burned the **Lady Johnson II**, out every spring while the seal fishery lasted and even invested in another motor vessel for the same purpose, the **Clayton M. Johnson**. In the meantime, Carino Co., a subsidiary of G.C. Rieber & Co. of Norway, opened a processing plant in Dildo, Trinity Bay, where its own ships and other Canadian ships sold their raw pelts. Carino also encouraged the sealers, especially the landsmen along the northeast coast, to concentrate on the quality of the pelts and the condition of the skin. Consequently, in the early 1980s, just as the sealers were beginning to receive a better price for their top quality pelts, the sealing industry was coming to an end due to the closure of the European markets to imports of whitecoat skins.

Pictured here is a sampling of the captains and ships that were active in the seal fishery during its final decades between 1940 and 1980.

Captain Morrissey Johnson (1930-). Born Catalina, Bonavista Bay, Newfoundland. Businessman. Captain of the MV **Lady Johnson**. Elected Member of Parliament, District of Bonavista, Trinity, Conception, 1984.

Captain Guy Earle (1917-1968). Born Carbonear, Newfoundland. Businessman. Part-owner and Captain of the SS **Kyle**, 1960-1965.

Captain Sidney Hill was born 4 November 1888 at Wesleyville, Bonavista Bay. He married Miss Sophie Winsor and they had two sons and three daughters. He first went to the seal fishery with Captain W.C. Winsor, Edward Bishop, and Abram Kean. In 1933 Captain Hill was given command of the SS **Eagle** and continued to command sealing steamers until 1941, serving as master of the **Beothic**, the **Imogene, and the Neptune**. In 1938, as captain of the **Beothic**, he brought in 42,156 pelts and his men received $128.81 each. In the 1940s and '50s Captain Hill commanded a number of motor vessels, including the **Arctic Sealer** and the **Arctic Prowler** for Shaw Steamships Ltd. of Halifax.

He died 3 September 1961.

83

Because there is no longer a significant market for seal products, the harp seal herds are not currently a valuable resource for Newfoundland. The herd, though diminished since the eighteenth century, has remained substantial and during the last years of the industry Newfoundlanders harvested about 80,000 to 100,000 seals annually, a figure similar to the number harvested at the beginning of the nineteenth century. Although landsmen continue to hunt adult seals close to shore, there is no active hunt on the whelping ice.

Index

Bibliography

Brown, Cassie, with Harold Norwood. *Death on the Ice*. Toronto: Doubleday, 1972

Chafe, Levi. *History of the Newfoundland Seal Fishery from the Earliest Available Records down to and Including the Voyage of 1923*. 3rd ed. St. John's: Trade Printers and Publishers Limited, 1923. (The first and second editions, published in 1895 and 1905, were also used.)

England, George Allan. *The Greatest Hunt in the World*. 3rd ed. Canada: Tundra Books Inc., 1975. (Originally published as *Vikings of the Ice*, 1924.)

Greene, Major H.W. *The Wooden Walls Among the Ice Floes: Telling the Romance of the Newfoundland Seal Fishery*. London: Hutchinson and Co. Ltd., 1933.

Ryan, Shannon. *The Seal and Labrador Cod Fisheries of Newfoundland*. XXVI. Ottawa: National Film Board of Canada, 1978.

Ryan, Shannon, "Seals Spelled Survival." *Horizon Canada*, III, No. 27 (August 1985), 638-43.

Ryan, Shannon and Larry Small, eds. *Haulin' Rope & Gaff: Songs and Poetry in the History of the Newfoundland Seal Fishery*. St. John's: Breakwater Books, 1978.

SRMD 415. Mr. Fred Bragg. Interview by Martha Drake, December 2, 1986. Shannon Ryan's Oral History Collection.